Blue Sky Dream

DAVID BEERS

Blue Sky Dream

A Memoir of America's Fall from Grace

A Harvest Book
Harcourt Brace & Company
San Diego New York London

Library of Congress Cataloging-in-Publication Data
Blue Sky Dream: a memoir of America's fall from grace/David Beers.
p. cm.—(A Harvest book)
Originally published: New York: Doubleday, 1996.
ISBN 0-15-600531-X
1. Beers, David. 2. Aerospace industries—California.
3. California—Biography. I. Title.
[CT275.B5562A3 1997]
338.4'762913'092—dc21 97-25493
[B]

Printed in the United States of America
First Harvest edition 1997
A C E D B

For my family

ACKNOWLEDGMENTS

The unusual form of this book leaves me indebted to three kinds of persons: members of my family who not only gave support but entrusted me, in a sense, with their lives; colleagues who shared insights as I wrote; and authors of works about the milieus I describe. "A communal memoir" is the phrase my fine editor, Bill Thomas, uses for what has resulted. Well, quite a community (one so large that to identify every member is impossible) deserves credit here.

I say thank you, profoundly, to my father and mother for their openness to this project. My father likes to joke, "I wish when you were born I had noticed the label warning: 'Everything you say can and may be used against you.' " This book, however, would never have been attempted without the willing participation of Hal and Terry Beers, and never for a second have I taken for granted this gift from them. Thanks as well to my beloved sisters and brother, Marybeth MacLean, Maggie Beers, and Dan Beers. And an eternal thank you to Deirdre Kelly, my wife and best friend, whose intelligence and honesty and caressing wit have made this book (as with all things good in our lives) possible.

Richard Rodriguez encouraged me years ago, over many lunches, to think in terms of a memoir. That he who wrote so masterfully of his own formation decided to take an interest in mine remains an inspiration to me. I thank Molly Friedrich for her surehandedness as agent and advisor. I thank Bill Thomas for his perfect way with my variable psyche and prose. For giving me much of their time to help me make sense of the changing technological and social landscape, I owe these brilliant people: Gary Chapman of the 21st Century Project in Austin, Texas; Ann Markusen of the Project on Regional and Industrial Economics at Rutgers University; Lenny Siegel of the Pacific Studies Center in Mountain View, California; Peter Calthorpe of Calthorpe Associates, San Francisco, California. For supplying me invaluable help along the way, I am grateful to Kathryn Olney, James Glave, David Kirp, and Peter Neushul. For their companionship and unfailing interest in my sanity, I thank Bill Richardson and Wallace Robinson.

A heartfelt thank you also to the gracious and creative people at Doubleday, especially Marly Rusoff, Sandee Yuen, Janet Hill, Jennifer Daddio, and most especially Jacqueline LaPierre.

While it is too daunting to cite every article and book informing what I have written, I do wish to list these most useful road maps. (Any wrong turns I may have taken are my own responsibility, of course.) *Barons of the Sky* by Wayne Biddle, for aircraft industry history; *Arming the Heavens* by Jack Manno, for early aerospace and Wernher von Braun; . . . *The Heavens and the Earth: A Political History of the Space Age* by Walter A. McDougall, for strategies of Cold War leaders and spy satellite history; "The Genesis of Silicon Valley" by Annalee Saxenian and *The New Book of California Tomorrow* edited by John Hart, for the development of Santa Clara Valley; *Crabgrass Frontier* by Kenneth T. Jackson and *Redesigning the American Dream* by Dolores Hayden, for the form, function, and growth of suburbs; *The Rise of the Gunbelt* by Ann Markusen et al., for the impetus behind aerospace settlements; *The New Alchemists* by Dirk Hanson and *Fire in the Valley* by Paul Frieberger and Michael Swaine, for the history of Steve Wozniak and the personal computer revolution; *Behind the Silicon Curtain* by Dennis Hayes, for the culture of Silicon Valley's boom.

Portions of this book have been published in different form in *Harper's, Mother Jones,* and the magazine of the *San Francisco Examiner.* To protect privacy, the "Gianninis" and the "O'Mearas" of Chapter 5 are pseudonyms.

CONTENTS

PROLOGUE

TAIL-END CHARLIE

On the wall of a hallway in the stucco four-bedroom ranchette where I grew up, the home of my mother and father, there hangs a family totem typical of our tribe: a neat grouping of twenty-seven photographs. The style of expression may well strike you as naive, even primitive, but don't be fooled. Beneath the crude Kodak colors lie wry paradoxes, ominous contradictions, the dense mythology of a people who believed themselves without myth.

Some help, then, in deciphering.

Note, first of all, that just about every one of these pictographs is an homage to one of us, the four children. We are immaculately backhanding tennis volleys. We are holding blue ribbons as we stand grinning in Speedo swimsuits. We are laughing without front teeth under First Communion veils. We are wearing scholars' robes over surfers' shorts as we reach for our many college diplomas. We have been told by department store photographers to stand all together just so, to radiate relaxed love, and without fail every time we have done it.

We know how to make it look easy, don't we? *A tribe's teaching, instilled early.*

Do you see how comfortably familiar we find the glare of sunshine, even as it seems to punish those other people, the visiting relatives from places strange and ancient with names like Rock Island, Illinois? Look how they stand there next to us, squinting and glistening in their stiff, wrong clothes. Sunshine burns its presence into every one of these pictures, even into those holiday snaps taken at night around the dinner table, because in every picture the skin of us, the four children, is freckled or browned or rosily peeling or all of these mottled together by our daily doses of ultraviolet. *A tribe's markings, read as blessings.*

And yet, in no picture is the light so bright as when it washes over the two people in this one portrait, over here, hanging just around the hallway corner. These two people, off by themselves, are my mother and father when they were simply Terry and Hal, on their wedding day.

As you see, this pictograph is unlike the others in that it is black and white with the lush, deep tones of a movie still. It was taken on April 21, 1956, in Corpus Christi, Texas, a place we, the four children, have never seen.

What we've been told of Corpus Christi is that back then it was a place very much of its time. It was one of any number of Sunbelt way stations where young men and women, fleeing flat prospects in cold climates, would find themselves for a brief while, would meet on a beach or at a party, would tend to fall quickly in love, and a few months later, without a lot of agonizing, would be married. So united, the two would hurry on to even more temperate climes and lives constructed around the promise of us, the four children—all of this accomplished with the grateful participation of the United States government. *A tribe's legend, taken on faith.*

As you see, the artificial light in this wedding picture has made my mother's dress, white lace and chiffon, all the more white, setting off her slender tanned arms and full black hair and ivory smile.

As you see, too, the light has made my father's uniform, the whites of a Lieutenant Junior Grade, all the more white, setting off his lean tanned face, his own full black hair, his own ivory smile.

The cake is bright white, the table cloth is bright white. All that is behind is plunged by the powerful flash into blackness. Into invisibility.

My mother, notice, looks just the slightest bit shy, as if she might suddenly lower overwhelmed eyes. Not my father. He looks proud. He is proud of his new bride and of how his life appears to be forming at age twenty-three. There is something else, I happen to know, that makes the groom, Hal before he was my father, stare straight into the burst of light so assuredly on this day. He is particularly proud to be wearing, pinned to his white officer's uniform, the gold wings of a naval aviator. *A tribe's talisman, sign of the elect.*

What did it mean to be a twenty-three-year-old naval aviator in 1956? This you will need to know should you want to give the pictographs their deeper reading. In answer, I can do no better than relate the tale my father tells, an adventure story. His tale concerns the time he was selected to set a speed record in the fighter jet the nation had just given him and taught him to fly, the F9F-8 Cougar.

At the time, jets were a recent enough phenomenon that the F9F-8 Cougar could trace a direct lineage to one of the very earliest. As the Second World War closed, the engineers of the Grumman Aircraft Engineering Corporation had borrowed from the catalog of standard propeller plane shapes—triangle tail, bullet nose, wings sticking out from the fuselage at right angles—and into this collage they had tried mixing the recently invented jet engine. Their product, named the F9F-2 Panther, had proven workable enough that the Navy bought some, and so, over the next decade, the engineers tinkered with their design, adding power, fuel capacity, eventually getting around to sweeping the wings back in such a way as to inspire boys and Detroit automakers. By 1956, their evolved result, Panther metamorphosed into sleeker Cougar, had the look of the jet age thoroughly arrived.

Of the few hundred Cougars in existence, my father's squadron took delivery of fourteen. Almost immediately, the squadron commanding officer, whose name was Jerry Robinson, perceived opportunity. In the new jet's improved ability to cover swaths of ocean, outrun the enemy, and return to ship, Jerry Robinson saw the speed and range needed to do something else, to break a record. The F9F-8 Cougar, he realized, was capable of flying across North America and back again in less time than any aircraft before. Here, then, was a fact of progress waiting to be established.

Jerry Robinson, as my father remembers him, was the sort of flying cowboy Hollywood gives us, the maverick ace with the sly wink, a well-liked leader "full of shady ideas. He never could go about anything straight ahead." That an Air Force pilot owned the standing record for a coast-to-coast-to-coast flight made Naval Commander Robinson hunger for it all the more. That the higher Navy brass, if it got involved, might want to throw a loop of red tape around Robinson, even pull him off of the attempt, caused Robinson to develop one of his shady ideas. He kept quiet his try at the record, asking no permission of superiors, pretending instead that this was just another training exercise for his men, well within his authority to direct.

Of course, there was nothing at all routine about it, and all involved knew it. To fly a San Diego–Long Island round-trip under eleven hours and twenty-seven minutes, as was the goal, would require keeping refueling stops to a minimum. Robinson therefore demanded that every squadron member devise the most efficient possible flight plan for the attempt, after which he picked the route he liked best and divided his men into teams, sending them ahead to far-flung airfields where, like pit crews, they stood ready at the gas pumps. On October 5, 1956, well before dawn, Robinson broke routine by having his F9F-8 Cougars towed to runway's end, engines dead to conserve fuel. He bent protocol further by ordering the jets' tanks topped off by hand because, at that hour, the night chill contracted the JP Kerosene fuel, maximizing what could be pumped aboard. Then, to conserve still more fuel, he treated the usually sacrosanct preflight safety check as a brief nuisance, idling his engine a mere thirty seconds before shoving the throttle forward and heading into the sky. Scrambling up to join him were three members of his squadron, tapped by Robinson to make the run with him as insurance. The last of these was my father, flying, as he likes to say when he tells his story, "tail-end Charlie."

My father remembers leaving behind the lights of Miramar Naval Air Station and San Diego, climbing into "inky blackness," cruising on in blind solitude until, on ahead, the contrails of his air mates gradually painted themselves against the sunrise. He remembers the four of them touching down two and a half hours later in Olathe, Kansas, and finding, on that tarmac carved from cornfields, newspaper people who'd been alerted by Robinson's advance team that some sort of history was in the making, some fact of progress was being established. He remembers Navy whitehats scurrying to refill his plane. He remembers, then, Jerry Robinson giving an order that was nonroutine in the extreme: Take off downwind. Doing so might conserve yet an extra few pounds more of jet fuel. But downwind was also the direction that would require the most runway for my father's F9F-8 Cougar, now at its heaviest and most combustible, to achieve flying airspeed. Given the wind conditions, it was possible my father might not be able to get off the runway in time, and if so, everything would end in a JP Kerosene fireball.

"I remember vividly hammering down that runway," is the way my father's telling continues, "watching the barbed wire fence at the end coming closer, closer, closer. And about fifteen hundred feet or so before the end of the runway I'm still on the pavement. The fence is closing rapidly and I'm still on the pavement, and I don't dare rotate for climb because I haven't the speed to dare drop the flaps yet. So. At what I perceive to be just as late as I could lower the flaps, I hit the flaps lever, the flaps come down, and I rotate the airplane. And I have this vivid recollection of watching that barbed wire fence go underneath the airplane. I can *see* the barbs on the barbed wire, no problem. Fortunately, Olathe is in the middle of a plain, no hills to get over. So once airborne, you stayed that way. But it was a close thing."

What did it mean to be a twenty-three-year-old naval aviator in 1956? For my father (as he told me in one of our many discussions now that he's had forty years to place things in perspective) it meant to have discovered "the one thing I was good at that no one told me to do. No one told me to fly. I liked it. I was good at it. And it was my own idea."

True, no one had told my father, whose own father was a man of sternly conventional expectations, that he must be a jet fighter pilot. But he had been asked, in a sense. He had been asked by the World War II propaganda he absorbed growing up, by the movie house cartoon called *Victory Through Air Power* animated by Disney, by comic strip air heroes like Milton Caniff's *Terry and the Pirates*. In high school, it was noted that he was the sort of teenager who was smart with numbers and painted model airplanes and had serious thoughts about the sky and planets and such, and so my father was asked by an ROTC naval officer if the people of the United States might finance his college education with a scholarship. Having educated the teenager in aeronautical engineering, the United States government asked if he wished mastery over the very machines the United States government had caused the teenager to dream about. America called this boy's adventure a nation's much-needed work and, for answering yes, rewarded my father with what an unshaped man craves: a reason to take himself seriously. The Navy, America, gave young

Hal rank, salutes from men twenty years older, dinners in the Officers Club served by black-skinned mess men, gold wings, an elaborate and ready-made construct of self-regard.

It gave him, most lastingly, a language to speak, the language that America was teaching its promising boys back then, a technicalese that seemed capable of saying everything necessary for the job of developing a postwar empire. It was a language languidly droll about what might electrify the soul: *I was tail-end Charlie. I rotated the airplane.* It was a language meant mostly, in its flat precision, for talking about problems solvable, the control of variables, the elimination of surprises. *Plan the flight. Fly the plan,* the Navy, America, taught my father to say with conviction. My father loved that language, mastered it with a convert's gusto, searing it onto the goofy tongue of the teenager until the new talk was all that was left, fused so thoroughly with his personality that it still inflects his everyday sentences. In my father's parlance, even if you're just fixing a toaster, you don't "try this or that," you *exhaust the various possibilities*. You don't "figure it out," you *troubleshoot successfully*. And when it's time to do the next doable thing, you don't "get around to it," you *turn to*. Words of crisp, controlled action, a language built on the notion of worth proven through competence. A language earned.

I liked it. I was good at it. And it was my own idea. If you proved competent enough, the place where the Navy trained you to fly fighter jets was Corpus Christi. We are told Terry happened to have moved there because tropical Corpus Christi was not at all the Midwest of her girlhood, and that she was a medical technologist at Driscoll Children's Hospital. We are told that a Navy friend of Hal's happened to chat up a roommate of Terry's sunbathing on the local jetty, and that as a result Terry happened to end up wearing a homemade sarong at one of the luau parties the naval aviators held around the Officers Club swimming pool. We are told Terry happened to meet Hal there, and then saw him again at one of the parties thrown by Terry and her friends. These were parties where, by the end of the night, the two parakeets named Hit and Miss invariably would have been let out of their cages and then one of the men who tended to be invited—they were oil geologists and interning doctors and military aviators, the better class of Sunbelt transients—would look funny as he tried to catch the birds and put them back.

"Your father liked to laugh," Terry says of Hal now when asked why, six weeks after their first date, she agreed to spend her life with him. "We both liked to laugh. That was very important. And when he came over for Thanksgiving dinner he did the dishes. I noticed he was very polite and I sure liked that." This is how my mother and father talk about Terry and Hal back then, in phrases that never focus to a precise dot, with the broad positives that attach to the goodly characters in folktales. We, the four children, nod in smiling wonder since we have no way to know if it was anything different.

By summer the newlyweds were finished with Corpus Christi and living in a tiny apartment in San Diego. My father had been assigned to a fighter squadron based at Miramar Naval Air Station, and there he was given his F9F-8 Cougar to fly. It wasn't the fastest jet in the world (though it cruised at Mach .86, about ten miles a minute), but what made the F9F-8 an advance for the day was its range, the fact that its wings had been equipped with extra fuel tanks. So, its wings engorged with fuel, its body an air-rammed blast furnace, its pilot breathing bottled oxygen inside a Plexiglas bubble, the F9F-8 Cougar was very much what a jet was made to be: a machine that consumed ferocious amounts of energy while wildly shrinking time and space.

What this meant to my father is that he could phone his parents from San Diego as he did one summer morning in 1956, casually announce he was coming to visit them, burn three tons of JP Kerosene on his way to Seattle, roar low over the air base before landing and, wearing a helmet with six stars painted across it, greet his awed mother and father in his fluent new language: *Made good time. Airplane performed well. How's everyone?* What could they say to him? He was telling them America's favorite story, the story of the son who zooms past his father's sternly conventional expectations, the son who roams far and returns transformed. The son in his F9F-8 Cougar had made himself such an exotic stranger, so much his own man, that he could roam in and out of his parents' lives at a speed of ten miles a minute.

Such a fantastic thing as the F9F-8 existed, of course, because there had been a Second World War, which rescued a commercially faltering U.S. aircraft industry with government contracts. The contracts paid for the building of a third of a million aircraft, increasing the industry's annual output by 13,500 percent while generating breathtaking technological advances and profits.

The F9F-8 existed, too, because there followed a Cold War, which again rescued a commercially faltering U.S. aircraft industry with, again, government contracts. The F9F-8 existed because America had been made to believe that there must forever after be a great and dominant aircraft industry subsidized by the nation's citizenry. America was informed of this as early as the waning months of 1947, when President Truman's Air Policy Commission summoned 150 men to give their testimony, virtually all of these men warriors, politicians, or industrialists with much to gain from an aviation boom. The men were in remarkable agreement about what was finally said in the commission report, titled *Survival in the Air Age*. It said that "This country, if it is to have even relative security, must be ready for . . . a possible World War III." It said that this World War III would be won or lost in the air. It said that a massive air armada therefore should be purchased by the U.S. government over the next four years. This view drew enthusiastic praise from *The Wall Street Journal* and from *Businessweek* and *Fortune* magazines. For anyone else worried such spending might ruin the economy with inflation, *Survival* offered this answer: "Self-preservation comes ahead of the economy."

In other words, if what was good for GM was good for America, here was a different formulation for a different industry, a formulation we would hear over and over again for the next half century. What was good for the aircraft industry was essential for the very survival of Americans. The making of flying machines was not to be seen as any mere business. It was to be imagined as a project of the nation's collective will.

Though Truman hadn't near the money to do all that *Survival* urged, in the spring of 1948 he did increase Pentagon aircraft spending 60 percent. And in 1950 when the Korean War provided a handy harbinger of World War III, the aircraft industry swung into robust revival. From 1947 to 1951, the aircraft workforce nearly doubled; by 1957, it had doubled again to nearly a million, well eclipsing the auto industry. Steeply increasing, too, was the money spent for aircraft research and development, with nine out of ten dollars donated by the U.S. taxpayer. The more complex flying weapons became, the more their makers spoke the new language of technicalese, so that by the mid-1950s the aircraft business was a sign of workforces to come, a manufacturing industry in which blue collars were a declining minority. To look down on the design bays of companies like the Grumman Aircraft Engineering Corporation was to view acres of white shirts, bright young men bent over row upon row of drafting boards as they cleaned up shapes for World War III, shapes like the F9F-8 Cougar.

To be a twenty-three-year-old naval aviator in 1956, then, meant having the good fortune to be handed the keys to what the richest nation in the world, its resources mobilized for war, had made its number one technological priority—the jet airplane.

It was all the better fortune that this was a moment when there was no "hot" war to fight. And so, over the Chocolate Mountains of California's southern desert, my father would pepper full of holes a polyester banner towed by a comrade, and over the Philippine Islands he would turn his guns on sea-washed rocks, occasionally scaring the fisherfolk. The four 20-millimeter cannons of the F9F-8 Cougar were toys: "Target practice was fun, a game,'' remembers my father. Yet he could feel part of a swashbuckling tradition because he was flying with war heroes like Jerry Robinson, the commander who ''never did anything straight,'' who had flown a small float plane off battleships, landing in the open sea to pick up downed U.S. pilots during World War II.

What Jerry Robinson appreciated in young Hal, though, was not heroism but a certain straightness, a personal philosophy, in my father's words, that ''flying was to be approached as a science.'' Jerry Robinson noticed that my father had taken the initiative to work out the F9F-8 Cougar's fuel consumption rates and make charts for all the squadron to strap to their knees when in the cockpit. It was an ''aero-weenie'' thing to do, says my father now, but it was something his flight commander saw and liked, and that, probably, is why out of a group of twenty flying officers, Jerry Robinson had chosen my father to join his run at a speed record.

With good fortune so easily wafting his way, there seemed to be no hurry to think creatively about life past one more day aloft in the F9F-8 Cougar. There were not yet any air combat veterans over Vietnam, though a squadron mate would go on to be one. There were not yet any astronauts, though a fellow naval aviator, John Glenn, would become the first American to orbit Earth. Perhaps Hal would remain in the Navy and be another Jerry Robinson. Perhaps he would move to the controls of the coming commercial jet airliners. If my father had any favorite notion of his fate, it was that he would live the aero-weenie's ideal as a troubleshooting test pilot. Anyway, why worry about it? The options all seemed so good and readily in view and directly in his flight path.

Here is how a cumulus cloud is formed. A bubble of warm air meets cold and up it rises, cooling as it expands, eventually reaching dew point and condensing its moisture into steam, a process that releases heat and causes the cloud to rise, releasing more moisture and lifting it all the higher. The aero-economy of the Cold War was beginning to be like this. A bubble of government money sucked away from other places and warmed by competition with the Soviet enemy had become a cloud of self-perpetuating steam enfolding millions of lives and livelihoods. That billowing economy had created the F9F-8 and paid for my father to be in its cockpit, had allowed my father to believe that he landed in that cockpit by just being himself. *I liked it. I was good at it. And it was my own idea.*

This, then, is what it meant to be a twenty-three-year-old naval aviator in 1956 if you were my father. It meant to consider oneself a fact of progress waiting to be established. Hadn't Jerry Robinson seen that in young Hal, just as he'd seen it in the new F9F-8 Cougar? Wasn't that why my father had been selected to be here, safely off from Olathe, Kansas, buoyed by an updraft of national will to an altitude of 40,000 feet, an altitude at which visibility appeared unlimited?

"So away we go," my father tells it, "the four of us, and we're in a little tighter formation this time, but I'm basically alone—it's not like you're snug up against another airplane. And there's Chicago over there to the left. Cleveland's up ahead. Louisville is down there to the right. It's just a gorgeous day. You could see forever. And as I'm gazing around at the scenery I look over at my left wing and I notice the trim tab is deflected quite a bit. Now, the trim tab on your wing trailing edge is meant to correct any left wing/right wing heaviness. That is, if one wing has more fuel in it than the other, that wing will be heavy. And the way you correct for that is you deflect the trim tab a little bit to pick that wing up. You do this with a button under your thumb on the control stick, and you manipulate that button almost unconsciously, reflexively, whenever you feel a pressure under your thumb, the pressure caused by a wing dropping. Turns out I had been clicking the button over and over without even realizing it, until the left wing trim tab was very much deflected. So something was going wrong with the way fuel was fed from the left wing.

"Well, I finally deduced that my leading edge fuel tank was not being emptied by its electric pump. I made a quick calculation and realized I had enough fuel to make Peconic field, our turnaround point on Long Island. But in no way, without that leading edge fuel, could I hope to make the return trip, east to west, because of the prevailing winds. So I radioed ahead to Commander Robinson. I said, 'My leading edge fuel tank is not feeding, and we have to do something about that when we land at Peconic, and we have to do it quickly because we don't want to waste a lot of time on maintenance.' So. When we landed at the Grumman works at Peconic, we taxied up into a swarm of ground people. The Grumman ground people were all over my airplane, troubleshooting what had gone wrong and how to fix it."

But the ground people swarming below him could not fix my father's problem, not fast enough, anyway. And so he was left behind with the ground people, watching the three other naval aviators take off and head back to San Diego, where they would arrive in time to break the record by thirty-eight minutes, where their portraits would appear over newspapers' accounts about how they had established another fact of progress.

"I'm not going to get to do this one," is what my father remembers thinking as he stood on the runway with the ground people. "I'm not going to get to do this one. But hell, there will be another chance, something just as good." He did not guess that in thirteen months he would leave the Navy, and not to fly airliners or to be a test pilot, but to sit at a metal desk and wear the clip-on badge and the white shirt of an aeronautical engineer. He did not guess that the expanding cloud below his wings would catch up with him just enough to draw him into it, that America would stop asking him to be a jet pilot and ask him instead to be content with corporate life. He did not have any idea that he would ease into the easy thing, the thing that America was easing into, life in the booming Cold War industrial project. And that once he did, he would chafe against that choice for the rest of his employed life.

You should know, if you are to give our family totem its deeper reading, that my father's life as an organization man perfectly traced the arc of the Cold War aerospace industry. And that (much like America itself) he became more and more dubious about the deal he had made. What our colorful pictographs do not reveal is that our father was the polite man that liked to laugh, yes, just as Terry remembers Hal. He was also the man who carried within himself a mercurial, mysterious anger that wouldn't subside until Cold War America was finished with my father, and he with it. You should know, since we in our pictographs are so good at making it look easy, that life in a four-bedroom ranchette in the sun could be, in certain unspoken ways, an uneasy life. *A tribe's secret, guarded closely.*

You should know also that in spite of this, the tanned faces in the pictographs are smiling because all along, we—Hal and Terry and the four children—imagined ourselves to be living the inevitable future. And that we are very surprised today to discover we were but a strange and aberrant moment that is now receding into history. *A tribe's fall from dominance.*

My father, as he watched his fellow aviators fly west without him on October 5, 1956, could not possibly know any of this was ahead of him, because, after all, he was still the twenty-three-year-old naval aviator who wears the white uniform and gold wings in the wedding photograph. As you see, the man in that photograph, the groom, Hal before he was my father, is staring directly into artificial light so bright that it makes him and his new wife luminescent and plunges all that is behind into invisible blackness.

ONE

LUCKY STAR

My father and I are together in the sky. I am eight years old and he has taken me up for the first time in the Piper Cherokee, a four-seat, single-prop, family sedan of an airplane he rents every once in a while. My father's idea is that we will fly over our house and get a look at it from a new vantage point, from high above on a clear summer morning. We will fly over at an agreed-upon time and my mother will stand in our backyard and wave, and we will wiggle our wings back at her to tell her we see her.

"Won't be long till we get there. Five, ten minutes. Have a look." My father's voice in my headphones cuts through the vibrating thrum of the airplane's engine. He smiles and points out the window, tracing a circle with the tip of his index finger as if marking an intersection on a road map. I follow his finger, look down and, suddenly, vaguely, I am terrified.

My fear has nothing to do with any lack of faith in my father's piloting. This morning has so far been one of his happy, whistling days, and I have been reassured by the words he shares with the man in the control tower—*cherokee eight six seven niner whiskey over*—the way men must speak to each other when they know exactly what they are doing.

What terrifies me is not even the fact of being here, so high in the sky. Many times my father has told me look upwards, and we have stood together in the backyard watching some interesting silhouette pass overhead on its way to the nearby air base. Having been told so often to look into the sky, it seems perfectly natural to be here now.

No, the queasy doubt rushing over me comes from what is there below, staring up. Down there is a labyrinth of freshly scraped earth and cedar-shingled roofs and white cement driveways and severe little lawns and blacktop ribbons snaking in and out endlessly, every curve bent with such precise consideration that I feel I'm looking at some immense, far too complicated board game. The sensed design to it all is what vaguely terrifies. This is some stranger's unsentimental schema without our home, my family, me, at its center. Down there, we could be *anywhere*.

I search out the familiar shape of a cul-de-sac, for that is where we live, in the house at the end of our cul-de-sac. But what I see, like berries growing on trellised vines, are many, many cul-de-sacs. How could any one of these tiny, repeated shapes announce itself to be our very own? From up here one could never see the brand-new lemon yellow Naugahyde couch my family has just installed with such momentous pleasure in front of the television set. From up here one could not see my sister, Marybeth, four years old, or my baby brother, Dan, or the melon stomach of my mother, pregnant with my sister-to-be, Maggie. From up here one could not see my father's workbench cluttered with baby food jars filled with nuts and bolts, or the dusty model of the Grumman F9F-8 Cougar hanging over it by a nylon thread. From up here there is no way to see the sky blue walls of my bedroom.

A few years before, my father had made a down payment on a newly built four-bedroom tract house with enough yard, enough land, to be visible from the sky, and so today he naturally thought I'd be excited to come along with him for an aerial view of where we'd made a place for ourselves. But now I am looking hard and I cannot locate our lives amidst all the sameness below, and that is what terrifies me.

That is what terrifies, and instructs. A child absorbs such a vision and begins to sense, at some level, the imperative of making a bigger meaning of things. I wonder whether all these people living just like me might be *my* people—all of us, perhaps, with some shared story. I wonder whether the pattern below might be a mass of connections joining me to some whole.

My father banks the plane over to his side for a long, slow turn. He is looking hard, too. His chin is stretched forward, his eyes are scanning calmly, deliberately, but they are not yet coming to rest on any one thing. I am watching his eyes. And now they turn my way. And now my father gives me one of his winks. I am wondering if there might be, readable in the design below, some reason not to be vaguely terrified.

Sputnik was my lucky star, its appearance in the darkness a glimmering, beeping announcement that my family would not know want. As my mother and father have so often told it, on chilly nights in the fall of 1957 the two of them would stand on the back porch of a rented cottage in Springdale, Ohio, watching that first satellite arc across the evening sky while I, the firstborn son but a few weeks old, lay in my crib. Already my father was growing bored with testing jet engines for General Electric, his first job out of the Navy. But then, not long after the appearance of *Sputnik*, he received the phone call that set our lives in harmonious motion. The Lockheed Corporation's Missiles and Space Division invited him to come help build America's own satellites in a place that was said to be the balmiest, most fertile in all of California (and therefore the world), a place that was called, in the brochures Lockheed sent us, the Valley of Heart's Delight. This is how we came to live in our house at the end of the cul-de-sac.

This is how *Sputnik* came to be a lucky star to millions of others like us, for its distant pull caused vast sums of money to ebb from once favored states and to flood into America's outlands, causing tract home societies like ours to spring up in places like the Valley of Heart's Delight. *Sputnik*, a 23-inch aluminum sphere weighing 184 pounds, launched into orbit by the Soviets on October 4, 1957, sped the transformation of the aircraft industry into something called "aerospace." *Sputnik* did this by rewriting the rules of the Cold War arms race, and when it did, the new rules favored my people, our future.

The new rules emerged from one of the military's oldest and simplest theories: Occupy the high ground. Now the high ground was space. Now there would be a scramble to perfect the rocket boosters to lift us there, to build the space platforms from which we could track our enemies, perhaps fire down on our enemies, perhaps knock our enemies' space-borne weapons out of the sky. Now the money flowing into aerospace would be hundreds and hundreds of billions, and anything that was integral to aerospace, from supercomputers to science scholarships, would be shoved to the top of national budget priorities.

Now a generation of men like my father would spend their lives working on "projects" they could not tell their families about. Now their families would be imbued with a culture that made a faith of fashioning the fastest, farthest, highest technology. A culture that taught the more momentous a machine is, the less likely you are ever to touch it, perhaps even to see it, certainly ever to use it, because it is built to go *way out there*. A blue sky tribe whose fertile crescents were sunny suburbs like my own.

Here is an odd thing about this shared story of ours. My tribe's future was not anything that Dwight D. Eisenhower ever wanted. You would think it might be. He was, after all, the United States President whose sky *Sputnik* trespassed. And he was then the emblem of suburban gentility, the man who earned his good life, his famously eternal boy's grin of contentment, by making himself supremely useful to a militarized nation. Ike was the general who golfed. Wasn't that the very thing that we, my tribe, chose to see in the glint of *Sputnik?* We saw the promise of suburban gentility extended to us as we made ourselves useful to an aerospace militarized America. But no, Dwight D. Eisenhower didn't care much at all for our star or our story.

Ike didn't like us.

Eisenhower looked at the nation's accounts back in 1953, the year after he was first elected, the year the Korean War ended, and he was dismayed to see that Truman's steady rearming had stuffed the federal budget, a growing budget, nearly two-thirds full of defense-related spending. Eisenhower was a man who considered every extra dollar given to the Pentagon a threat to the dream of suburban gentility. He dourly joked that the Joint Chiefs of Staff "don't know much about fighting inflation." He said, "This country can choke itself to death piling up military expenditures just as surely as it can defeat itself by not spending enough for protection." And so by 1955 Ike was proud to declare he had cut the military budget by 20 percent. His desire to push that graph line further downward was merely bolstered the next year by his resounding re-election.

Like Truman, however, Eisenhower also firmly believed that America must continue to wage the Cold War. Also like Truman, perhaps even more so, Eisenhower saw nuclear weapons as the cheap way to go about it. If America could ring the Soviet domain with nuclear-armed bombers and convince the world we were ready to drop them on our enemy's advancing tanks and troops, then our citizenry need not choke on the cost of maintaining a much larger conventional military presence abroad. This approach explains why, during the very years that Eisenhower chipped away at military spending, he invested strongly in the foundation of aerospace. He did so not out of any zeal for military expansion, but in pursuit of a less expensive evil.

The inherent flaw to this strategy is that it worked too well. The Soviets did apparently come to believe the United States was willing to fight "strategically" with nuclear bombs, willing even to use our new, improved hydrogen bombs, a thousand times more powerful than atomic weapons before, as part of our official doctrine of "massive retaliation" against Soviet ground gains. And so the Soviets redoubled work on an H-bomb to answer our own and, having exploded one nine months after we did in August of 1953, they continued developing missiles meant to carry their H-bomb very far, as far as any golf course within any one of America's genteel suburbs. By the mid-1950s, both nations were pouring more and more millions into the perfection of intercontinental ballistic missiles.

Dwight D. Eisenhower dreaded this ICBM race not only because it was dangerous and expensive, but because it was creating before his eyes a new class he didn't like. In the late days of his presidency he would damn us with now infamous names, calling us "the scientific-technological elite" and "the military-industrial complex." He would warn of the "danger" posed if public policy fell "captive" to the culture of the technocrat, people like my father.

He would tell a group of reporters: "When you see almost every one of your magazines, no matter what they are advertising, has a picture of the Titan missile or the Atlas . . . there is . . . almost an insidious penetration of our own minds that the only thing this country is engaged in is weaponry and missiles." He would note that a weaponry-driven "technological revolution" made research more critical and also "more formalized, complex and costly," and that this meant "a steadily increasing share is conducted by, for, or at the direction of the Federal government." He would mourn what came with this technocracy: The state was now expected "to make life happy in a sort of cradle to grave security," and this augured the end of "self-dependence, self-confidence, courage, and readiness to take a risk." By the end of his presidency, Ike would say such things with the air of a general weary in retreat.

Yet, throughout his eight years in office, President Eisenhower found himself with little choice but to promote the ascendence of my blue sky tribe. Even his brightest hopes carried with them an implied lasting dependence upon us. Eisenhower knew his only bet for making ongoing military cuts was to stop the missile competition with the Soviets, to forge some arms control. This made one of the great obsessions of my tribe, the race into space orbit, a thing that Eisenhower, as much as he wished it, could not ignore. Should there be in the president's best scenario a treaty freezing the Cold War missile buildup, an orbiting satellite could verify compliance by looking down on the Soviets. Should there be instead Cold War as usual, a spy satellite could tell the president how many missiles were being added to the Soviet arsenal so that we might not over-invest in our own supply. Either way, a satellite could save money for the parsimonious Cold Warrior. Either way, the Soviets, with their improving ground-to-air missiles, were only getting better at shooting down our U-2 spy planes, trapped as they were in Earth's atmosphere. And so, in addition to the millions he forlornly gave the missile makers, Eisenhower signed the work orders for a fledgling satellite industry.

Once onboard, Eisenhower intended America's try at a satellite to be a thorough and successful one. But during most of his tenure, the satellite program was also largely civilian-directed and unhurried, for this president chose not to imagine that national prestige, or for that matter his popularity as a president, might be tied to an orbiting symbol. Eisenhower did not particularly believe in a space race. Not even America's stunned reaction to *Sputnik* changed that about Ike. *Sputnik,* he said, "does not rouse my apprehensions, not one iota . . . They have put one small ball in the air."

On November 7, 1957, four days after the Soviets had beaten us into orbit again with far larger *Sputnik II,* six tons of space vehicle carrying one dog, President Eisenhower told the nation:

> There is much more to science than its function in strengthening our defense, and much more to our de-

> fense than the part played by science. The peaceful contributions of science—to healing, to enriching life, to freeing the spirit—these are [its] most important products . . . And the spiritual powers of the nation—its underlying religious faith, its self-reliance, its capacity for intelligent sacrifice—these are the most important stones in any defense structure.

As I say, Ike did not care much at all for my family's star, our future, our story.

Perhaps that is why the blue sky children of my generation have made Dwight D. Eisenhower into a figure of addled docility. We have made him our retro mascot for a lost era's bliss, and this has allowed us to forget his scolds. He may never have liked us, but we like our idea of Ike. He is our kitschy joke of a grandpa.

Our true father is someone quite different, a man who not only embraced our star, our story, but wrote that story for us over and over again in whatever way suited the mood of the moment, in whatever way made our version of the future seem inevitable. Many times the ex-Nazi rocket scientist Wernher von Braun has been named "the father of America's space program." He is father to more. Wernher von Braun fashioned a new creation myth for his tribe and (lasting awhile, at least) for all of America.

Three years before *Sputnik*'s launch, this is what Wernher von Braun said about the prospect of a satellite to those American officials who held the purse strings. He said he could build a "humble" one right away for a mere $100,000, that it would work just fine, and that *"it would be a blow to U.S. prestige if we did not do it first."* In this familiar patter—*Low, low price. Immediate delivery. Be the first to have one.*—we find the hard-selling entrepreneur, a type that belongs to all America timelessly. But here, too, is a quality quite particular to my people, our moment.

The classic American entrepreneurial hero searches out unmet desires in the everyday world and then, with a certain flexible flair, invents the answers, products for the masses to use. Von Braun's genius lay elsewhere. He was brilliant at inventing new and different uses for the only product he ever desired to make, the space rocket. He was a master at selling his one product to the only customers who could ever afford it, a nation's rulers. The big, blond man who was gently funny and firm, always so firm in his purpose, had a certain flexible flair for making powerful men believe in him.

And because rulers survive by shaping and reflecting their nation's collective psyche, von Braun came to their aid in fomenting mass imagination, mass desire. Von Braun made it his business to understand the workings of so gossamer a thing as "national prestige" and what might be a "blow" to it. Wherever he happened to find himself, in Hitler's Germany or Ike's America, Wernher von Braun wrote the sales jingles for his product in the local dialect.

As a teenager, von Braun already was learning these skills of salesmanship from a mentor, the mathematician Hermann Oberth, famous for his pioneering calculations and speculations about space travel. In 1923 Oberth grabbed attention with a slim volume called *The Rocket into Planetary Space*, and in 1929 he published a greatly expanded version, *Roads to Space Travel*. In between, the beetle-browed professor with the black mustache signed with the film director Fritz Lang, who was looking for some spectacular way to publicize his upcoming space rocket movie, *The Girl in the Moon*. Oberth took studio money on the promise he would build the first honest spaceship, a two-stage manned rocket, and blast it off at the movie premiere. The professor, not surprisingly, was still tinkering with the very beginnings of his prototype well after the film premiere had come and gone.

But Oberth's vision had galvanized some young rocket enthusiasts in Berlin who formed themselves into the German Society for Space Travel, and one of these young men was Wernher von Braun. Eighteen years old in the year 1930, Wernher von

Braun came to Hermann Oberth and asked to work for free on the rocket motor that was left over from *The Girl in the Moon*.

In the Piper Cherokee we are flying, my father and I, somewhere between Earth and the other planets who look down upon us. I know the planets are there even when the sky is bright with daylight. I have been told this by my father, who always seems very pleased to be asked any question about the planets. I know that the planets will be there every hour of the day forever, and that the planets are connected to my father's work. To cover the empty wall of my new bedroom, my father has given me a map of the solar system, and so the planets are the first things I look at upon awakening. As I lay in bed studying their various faces made up of moon shadows and swirling gas patterns, Jupiter seems to me big and friendly, Saturn haughty with all its rings, Pluto sullen at finding itself frozen at the outer edge of outer space. The solar system is a crowded and important place with a rectangular border, I can see by the map, and a boy like me should know his way around it. Why else would my father have placed the solar system next to my bed?

After the end of the Second World War, Wernher von Braun, by then the world's leading authority on rocketry, came to live in the United States and began to sell his space rocket dreams, in many varieties of ways, to his new customer base. Upon arrival, he wasted no time dashing off a short fiction about going to Mars and meeting the green Martians who lived there. Over the years he summoned the biggest possible names, Charles Darwin and God, as endorsers for his cause. He said going to the moon meant "a completely new step in the evolution of man," and that "we are extending this God-given brain and these God-given hands to their outermost limits."

A good and early example of the supple appeal of Wernher von Braun lies in the article he contributed to a 1952 edition of *Collier's* magazine, a special issue titled: "Man Will Conquer Space *Soon.*" The Cold War was on, of course, and so von Braun's glossy prophecy began this way:

"Within the next 10 to 15 years, the earth will have a new companion in the skies, a man-made satellite that could either be the greatest force for peace ever devised, or one of the most terrible weapons of war—depending on who makes and controls it." Wernher von Braun saw market potential in the draconian logic of the Cold War.

Wernher von Braun saw potential, as well, in the decade's nervously conformist mood. And so, our "technicians in this space station . . . will keep under constant inspection every ocean, continent, country and city. Even small towns will be clearly visible," he promised. "Nothing will go unobserved." America might, in short, occupy the heavens to monitor deviance at home. We would conquer Communists and space itself while freezing in place what was clean and good and neighborly back on Earth. Those were just a few of Wernher von Braun's reasons for why his new country needed a space rocket.

The *Collier's* package included pieces by other space experts, but what made the parts join into an irresistible panorama were large, lush paintings by an artist named Chesley Bonestell. There was a swept-winged spaceship, more rakish than any Oldsmobile hood ornament. There was a white, windowed doughnut wheeling over blue earth. There was a desert mountain chiaroscuro and tiny human shapes in the shadows, gazing toward a red ball far off in the blackness. Under this last painting was the explanatory caption:

"Mars, at its closest 35,000,000 miles from Earth, as seen from the outer moon Deimos, where man could land before going on to the planet."

The words, postcard language, are of a perfect tone, for Chesley Bonestell was a painter of postcards from Wernher von Braun's inevitable future.

My father has shown me a book given to him as a boy that made him ache for that future. The book is *The Conquest of Space,* published in 1949 when my father was sixteen. The numbers-dense text is by one of von Braun's German peers, Willy Ley. The illustrations, dozens of them, are Chesley Bonestell's postcards from *out there.* On orange Venus, "windblown dust etching the rocks into fantastic shapes." On Mercury, mustardy mud plains cracked by sun so hot, "explorers could not leave the protection of their ship for long." On Saturn, rings of rainbow hues, viewed "during a midsummer night." On the moon, "the mountains of eternal light." On Jupiter, a fiery waterfall pouring out of the gloom; the "lake below is liquid ammonia." On Mars, a snowdrift foreground, just as it would appear while "looking towards the setting sun." Chesley Bonestell created the travel brochure for a sight-seeing trip through space, the collective vacation awaiting the nation when a generation's hard work was done.

As early as 1952 Wernher von Braun was telling America we could achieve all this and, with the shrewdest intuition about his latest client, he was conveying something more. We would domesticate space even as we carried on for the cowboys who were, at the time, cramming the television channels. We would do it even as we fulfilled Manifest Destiny: "Crossing the Last Frontier" is what Wernher von Braun named his *Collier's* article; the best-selling book that sprang from it he called *Across the Last Frontier.*

And yet nothing in this catalog of persuasions is the central story of my people, the story that makes Wernher von Braun our visionary patriarch. That story may be found in the introduction he penned for the 1962 edition of *The Mars Project,* a book he'd written a decade before. The language is not at all as sensual as a Chesley Bonestell painting. But beneath its beige tones lay a new creation myth for America, a re-imagined story about how technological superiority comes alive.

The old myth taught that the lightbulb, the telephone, the

airplane—almost every American history-changing machine—had sprung from a lonely visionary tinkering alone. First, individual triumph. Then, via the marketplace, national progress. But that was a quaint story now finished, and Wernher von Braun said as much. He wrote that we of the now-arrived space age must face the death of "backyard inventor, the heroic inventor."

Accept it, said Wernher von Braun. Any "true" space effort "can only be achieved by the coordinated might of scientists, technicians, and organizers belonging to very nearly every branch of modern science and industry. Astronomers, physicians, mathematicians, engineers, physicists, chemists, and test pilots are essential; but no less so are economists, businessmen, diplomats, and a host of others."

A *host* of people, Wernher von Braun saw, all ready to be *coordinated* and put to work on great projects of the state. The time had come, he was saying, for nothing less than the reorganization of America's way of making a living. Wernher von Braun saw, in missiles and satellites and spaceships and the people who would create them, no threat to suburban gentility. Unlike angst-ridden Ike, Wernher von Braun saw in aerospace the work of a worthy new middle class, not an elite so much as a *host* of scientific-technological-military-industrial-complex families living lives of modern purpose. He saw my family, joined to some bigger design.

A bit further down in the same book introduction, the father of our tribe discusses the "grand scale" of work to be done on his "flotilla of space vessels" destined for Mars. The crew would be "not less than 70 men." Presumably they would leave behind the millions more of us building his space rocket product, living in our many Valleys of Heart's Delight. Here is the central story of my tribe, as told by Wernher von Braun:

"Great numbers of professionals from many walks of life, trained to co-operate unfailingly, must be recruited. Such training will require years before each can fit his special ability into the pattern of the whole."

———

"Got your bearings yet?" My father is helping me locate myself in relation to the earth as we fly on toward our neighborhood, our house, my mother who will be down there waving to us. "Those big structures there, recognize them? They're the hangars at Moffet Field." My father has told me about those before, hollow hulks by the highway, once home to huge military blimps, somebody's idea organized on a grand scale that turned out, in the end, to be a detour on the way to a dead end.

"Next to those, in there somewhere, is your dad's place of work. That's Lockheed." This holds great interest for me, given that I have never been inside Lockheed, have only seen, once from the company parking lot, the barbed wire and chain-link and sentry kiosks that my father passes through whenever he goes to work. What I see, as I follow my father's eyes down, is a very large, white radar dish pointed into the heavens, and around it many tight-lidded shoe boxes, windowless, gray, some with smaller dishes and antennas twisting off their roofs. *In there, somewhere*. I remain vaguely terrified.

"See the freeway? Going in over there," says my father, pointing to a great trench of flattened dirt. I know there is a great trench of flattened dirt not far from our house, but I also know it is but one of many. Wherever my family drives, it seems I see out the car window the yellow graders and dump trucks and bulldozers making their endless gouges in the ground. And so I wonder, is that great trench of flattened dirt down there ours?

"Eighty-five, and over there two-eighty."

Yes, those sound like ours, the ones the adults talk about eagerly as they stand in front of their new tract homes on these warm summer evenings, their cigarettes glowing, their laughter drifting over to us kids as we chase each other from yard to yard. Eighty-five and two-eighty sound like the freeways that, if ever finished, will connect to the expressways that connect to the street that connects to our cul-de-sac.

It took little time at all for Lyndon Baines Johnson to exploit the luck of *Sputnik*. The senator from Texas had a problem, which was that a wing of his Democratic Party wanted the desegregation of America, and this probably made his party's next presidential candidate unelectable. What LBJ saw in Sputnik, therefore, was a fortunate icon of promise and peril to dangle above the middle class he needed. He had in hand a political consultant's memo saying the issue of *Sputnik* "is one which, if properly handled, would blast the Republicans out of the water, unify the Democratic party, and elect you president."

He opened proceedings of his Senate Inquiry into Satellite and Missile Programs on November 25, 1957, by describing the two *Sputniks* as a technological Pearl Harbor that Eisenhower's Republicans had missed on their watch. He, for one, was worried about U.S. missile supremacy. He summoned Vannevar Bush, the framer of "big science" policy for Franklin Roosevelt, and Vannevar Bush declared *Sputnik* "one of the finest things Russia ever did for us" because it "has waked this country up" to the need for more science education, more respect for the scientist "as a fellow worker for the good of the country."

Americans listened to all this and pondered in their newspapers a certain proof that Communists were heartless. To great outrage here, the dog in *Sputnik II* had been left, by Soviet plan, to die in space.

This is the point in time where Walter A. McDougall, noted historian of aerospace politics, locates the full firming up and taking charge of what he calls the space era "command economy." The command economy was the sort of economy Dwight D. Eisenhower dreaded and Wernher von Braun made his creation myth. It was the government mobilizing and funding technology with a zeal that would burn even hotter once LBJ and John F. Kennedy rode their *Sputnik* scare strategy to the White House in 1960.

It was the state commanding, in a sense, that not only technology but *places* come into being.

Blue sky metropolises, nurtured by federal dollars, would be commanded to rise out of orange groves and scrublands and prairies and deserts and other former boondocks to industrial America. The money would flow to the Northwest of Boeing, to the Texas of Bell Aviation and Mission Control, to the Rocky Mountains of the North American Air Defense Command, to the Florida of Martin-Marietta and Cape Canaveral, to the Alabama of Wernher von Braun's U.S. Army Redstone rocket works. The money would generally flow to where land was cheap, even better if the land was federally owned and so could be bought at subsidized bargains. The money would tend to flow to places the military liked, and this often meant places wide open, remote, and quite far from the stodgy East.

The money would go, as well, to certain centers of technological innovation, the realms, for example, of Boston's MIT and Pasadena's CalTech, universities that made a specialty of military contracts. A basic rule of thumb is that the money flowed to where life could be made affordably good for the blue sky professional and family. This meant that even when the money found its way to Massachusetts and New York, it tended to pass up the old manufacturing cores for suburbias like Bethpage, Long Island (Grumman), and those around Boston's Route 128 (aerospace electronics).

Ann Markusen, a Rutgers professor of urban planning and policy development who has studied the patterns closely, has written that after the Korean War the flow of military contract dollars shifted away from traditional industrial centers in the Middle Atlantic and Northeast states, irrigating instead the Pacific Coast and "all other outlying regions." Looking at the population booms that resulted, she traces "an arc stretching from Seattle down through California and the Intermountain West, Texas, isolated spots in the Southeast and up the Eastern Seaboard to Long Island and New England." A great curve of blue sky communities, commanded into being.

The biggest of them all, of course, was and is Southern California, home to Aerojet and Convair and Ford Aerospace and Hughes and Litton and Lear Siegler and McDonnell-Douglas and Northrop and Rockwell and RAND and TRW and the United States Air Force Space Division and all the hundreds of subcontractors that serve them as well as many key military bases and universities. The San Fernando Valley of Southern California is home as well to Lockheed, and a time came in the late 1950s when Lockheed was commanded to build an answer to *Sputnik*. This caused Lockheed to seek out a second California home farther north and to locate its Missiles and Space Division in the Valley of Heart's Delight. Lockheed moved 2,000 families there within a few weeks in 1956, at the time the largest single move of corporate personnel. Over the next five years the division's ranks would rise to nearly 20,000, making Lockheed the area's single biggest employer by the early 1960s.

The Chamber of Commerce brochure meant to entice families like mine showed pictures of blossoming orchards in "California's All-Year Garden," one of "the most beautiful valleys in the world" where "life moves smoothly under a benevolent kindly sun."

We were assured, too, that a "prominent educational psychologist of Columbia University" had used "modern quantitative methods" in comparing "thirty-seven points of commonplace excellence" in 310 U.S. cities. His judgment of the Valley of Heart's Delight: "Tops in general goodness of living."

A similar brochure published in the mid-1960s offered data even more compelling: "People here feel more fit because they live more healthful lives," ran the copy, "and the average youngster is a few inches taller and a few pounds heavier than his counterpart in most other sections" of the country.

As a child, Wernher von Braun is said to have been possessed by the planets from the moment he first looked through a telescope.

The telescope was given to him when he was eight by his mother, a baroness (the re-imaginer of America's middle class was born of German nobility). The child Wernher von Braun came of age not long after the close of that first experiment of technological mass destruction, the First World War, and when, sometime around puberty, he read Hermann Oberth's *The Rocket into Interplanetary Space*, the boy found more than numbers. He found Oberth's suggestion, for example, that a nation's rulers might want to use immense, orbiting mirrors to focus the sun's rays onto an enemy's crops and cities, scorching them away like ants under a magnifying glass. Soon the teenager was setting off simple rockets outside of Berlin.

At age twenty Wernher von Braun met his most important mentor, General Walter Dornberger, who, in his plan to re-arm Germany, longed for a manned "boost-glide bomber" that could slip in and out of space, raining hellfire at will. The terms of the Treaty of Versailles, von Braun was told, were designed to quash German arms production—but those terms neglected to mention space rockets. They said nothing about gaining the ultimate high ground. General Dornberger and the German Army became Wernher von Braun's customers, and then, within five years, so did Adolf Hitler.

Much later, Wernher von Braun would say to America that any man who wanted to create a space rocket in the Germany of that era had little choice but to belong to the Nazi Party, that this is why he had been a registered Nazi. He would claim that he took money from the Nazi State early on thinking that "Hitler was still only a pompous fool with a Charlie Chaplin mustache." He would admit he saw Nazi guards forcing prisoners to build his rocket under gruesome conditions, and that he had heard some who resisted were executed. He would assure, however, that he "never ceased to be ashamed of the fact that even in a battered and gutted Germany struggling for survival such an outrage could have developed."

General Walter Dornberger, who headed the V-2 project, remembered different emotions. He remembered that Hitler had dreamed in his sleep that rockets would never be the weapons of terror Germany needed, and that the Führer considered this dream "infallible." Dornberger remembered how he and Wernher von Braun caused Hitler to dream differently. At noon on October 3, 1942, on a day when "the arch of a clear, cloudless sky extended over Northern Germany," Dornberger recalled standing with "Dr. von Braun" and various of Hitler's brass, watching the first successful firing of the V-2 "vengeance weapon." As the rocket "raced away at a speed of over 3000 mph," at the sight of that "tiny dot glittering dazzlingly white," one of Hitler's colonels wept with joy. And then, wrote the general, "like excited boys . . . everyone was shouting, laughing, leaping, dancing, and shaking hands."

The best at selling believe every one of their pitches, no matter how irreconcilable when laid side by side. Indeed, if the salesman Wernher von Braun was deft with the rational construct, he never let himself be captive to it. No, cool reason must at some point be put aside if one is to build a space rocket product, America was told in *Space Travel: A History*, a book cowritten by von Braun, which ends by reminding that "Henry the Navigator would have been hard put had he been requested to justify his actions on a rational basis."

Faith, as much or more than reason, caused space rockets to be made, Wernher von Braun instinctively knew. That clinging to faith probably allowed him to believe everything he ever said to every one of his potential customers. Erik Bergaust, a tireless promoter of early U.S. aerospace efforts, writes of his friend Wernher von Braun sitting by the campfire at night, lecturing his American companions on what God wants of men, on morality, on the "ethical laws" that are "enforced from upstairs." There is, in another book, a picture of Wernher von Braun's team standing smiling before the ruthless machine they have built for Hitler. Painted on the rocket is a winsome Girl in the Moon.

Taken together, all of the teachings of Wernher von Braun add up to nothing coherent, and yet they have offered his blue sky tribe, people like my family, an approach to life. Taking his example, we have found it possible to hold many profoundly contradictory notions in our minds as long as a sensation of forward momentum could be felt in our lives. A space rocket to preserve us, a space rocket to change everything, the evolution of animal man, the revelation of God's plan, the Girl in the Moon on the weapon of vengeance—Wernher von Braun gave his tribe any number of ways to explain to ourselves why we naturally deserved the fruits of an economy commanded from above, our new lives of aerospace suburban gentility.

When, along with dozens of other Nazi rocket experts, Wernher von Braun and Walter Dornberger became America's victory spoils, they dusted off visions of apocalypse for their new paymasters at the Pentagon. "I didn't come to this country to lose the Third World War. I lost two," Dornberger would tell the Air Force, still selling his hellfire raining space bomber. But it was von Braun's genius to hook into the American mass imagination not with destruction fantasies but with an immigrant's ode to his new country's can-do spirit. He stroked the ego of "the most fantastically progressive nation yet conceived and developed." We would assert our American exceptionalism with each Chesley Bonestell planet we visited. America would ride the optimism of its newly technocratic middle class into space and find even more optimism *out there*.

What did I, eight years old, know of Wernher von Braun as I flew in my father's rented Piper Cherokee in search of some shared story among the subdivisions below? I knew nothing specific, not even the name of Wernher von Braun. But I knew that on the television some mornings there was Walter Cronkite and a Mercury blastoff instead of cartoons, and I knew that my father was tied in some way to this, and that this might be why I lived in a place where people, like things in general, were always looking up.

"See the church there? The cross?" Yes, I saw it. "And over

there, those big letters, that's Shopwell.'' Yes, I saw our supermarket now. ''There it is, Dave, there's the house, juuuuuust about *right* below us.'' Yes, I was happily relieved to say, I could see the house—our house, our station wagon parked out front, our cul-de-sac, our backyard. I could see my mother, too, a speck marking our spot in the pattern of the whole, waving to us in the sky.

TWO

INVASION

Tony, Tony, listen, listen.
Hurry, hurry, something's missin'.

These are words my mother taught us for getting the attention of St. Anthony, who, she said, would guide us to whatever we were looking for but had not yet found. Hers was a perfect prayer for a blue sky family in the early 1960s, as colorfully casual as a tiki lantern, resistant to any doubt that we in our suburban frontier held the interest of heaven.

A crisis would develop. Mutterings, hard soles stepping hard somewhere in the back of the house, the *whooshing* sound of my father moving in his dark suit, moving with those quickened, long strides that sent us children edging into corners, up onto chairs, anywhere that was, as he would say to us, "*out* of the *WAY!*" My father's keys were missing again. He was yanking open drawers and shoving hands between seat cushions. He was muttering, "For cripes *sake.*" He was late for Lockheed.

Tony, Tony, listen, listen.
Hurry, hurry, something's missin'.

"It's worked before. You just have to believe," my mother would say, her voice upbeat. She would go to the sliding glass door, walk out onto the redwood deck, stand under the bamboo-thatched roof, move her lips in prayer just beyond my father's vortex.

Tony, Tony, look around.
Something's lost and must be found.

Soon enough someone, usually my mother, would be drawn to some unlikely spot, maybe to a clump of crabgrass near a Rainbird sprinkler. There would be the keys, waiting for my father's exasperated swipe at them. After my father and his keys had disappeared in a puff of exhaust around the corner, headed in the direction of Lockheed, we children would move out of the corners of the house, would reclaim the empty spaces for ourselves, and all the best possibilities for the day would be there for us, as if by some small miracle.

Some evenings my father would bring home to me new images for the filling of empty spaces, pictures to hang in my bedroom next to the solar system, publicity photographs of Lockheed products. There were stubby-winged jets and fire-swathed rockets, satellites that hung in space like tinfoil dragonflies. And my favorite, the Polaris. "The most beautiful missiles ever fired," a U.S. Navy Rear Admiral pronounced the nuclear-tipped A1X Polaris, having witnessed its successful submarine test on a summer day in 1960. The fully evolved, deployed Polaris, designed under the guidance of Wernher von Braun's friend and fellow former Nazi, Wolfgang Noggerath, was capable of traveling 2,400 nautical miles in a few minutes and delivering, from its elusively mobile launchpad, three separate warheads to a single target deep within the Soviet Union—facts no doubt beautiful to a nuclear warfare strategist. The Polaris was beautiful as well to a boy who thumbtacked its picture on his wall, a pure and universal shape if ever there was one, white and smooth, perfectly frozen above the convulsed ocean surface through which it had just burst. Lockheed always photographed its missiles headed up, never killing end down. As a child I didn't wonder what the Polaris was *for*. Perhaps once launched it just stuck there in the solar system's firmament like a dart in the ceiling. Maybe it metamorphosed into one of my father's pretty satellites with the glittery solar panels. That the Polaris was so obviously the future exploding out of the sea seemed reason enough to create it.

My mother gave me her own pictures, Catholic holy cards, Virgin Mary visitations, saints aglow, Christ baring His Sacred Heart while floating up in the clouds. And so airfoils and angel wings, blastoffs and holy ascensions, Our hovering Lady of Fatima, her cloaked contour so *aerodynamic*—all of these images, my father's and my mother's—mingled in my child's mind to form a coherent iconography. An empty space was not so hard a thing to fill up if you were determined to see in it what you wanted.

That, my mother and father will tell you, is how they remember their brand-new tract home in their brand-new subdivision: as a certain perfection of potentiality. Nowadays, when suburbia is often disparaged as a "crisis of place" cluttered with needless junk and diminished lives, it is worth considering that it was not suburbia's *stuff* that drew people like my parents to such lands in the first place, but the emptiness. A removed emptiness, made safe and ordered and affordable. An up-to-date emptiness, made precisely for us.

"We never looked at a used house," my father remembers of those days in the early 1960s when he and my mother went shopping for a home of their own in the Valley of Heart's Delight. "A used house simply did not interest us." Instead, they roved in search of balloons and bunting and the many billboards advertising *Low Interest! No Money Down!* to military veterans like my father. They would follow the signs to the model homes standing in empty fields and tour the empty floor plans and leave with notes carefully made about square footage and closet space. "We shopped for a new house," my father says, "the way you shopped for a car."

Whenever I think of the house they bought and the development surrounding it, the earliest images that come to mind are of an ascetic barrenness to the streets, the lots, the rooms. The snapshots confirm it: There I am with my new friends around a picnic table in the backyard, shirtless boys with mouths full of birthday cake, in the background nothing but unplanted dirt, a stripe of redwood fence, stucco and open sky. That was the emptiness being chased by thousands of other young families to similar backyards in various raw corners of the nation.

"Didn't the sterility scare the hell out of you?" I've asked my mother often. "Didn't you look around and wonder if you'd been stuck on a desert island?"

The questions never faze her. "We were thrilled to death. Not afraid at all. Everyone else was moving in at the same time as us. It was a whole new adventure for us. For everyone!"

Everyone was arriving with a sense of forward momentum joined. Everyone was taking courage from the sight of another orange moving van pulling in next door, a family just like us unloading pole lamps and cribs and Formica dining tables like our own, reflections of ourselves multiplying all around us in our new emptiness. Having been given the emptiness we longed for, there lay ahead the task of pouring meaning into the vacuum.

Listen, listen . . . look around . . . must be found.

We were blithe conquerors, my tribe. When we chose a new homeland, invaded a place, settled it, and made it over in our image, we did so with a smiling sense of our own inevitability. At first we would establish a few outposts—a Pentagon-funded research university, say, or a bomber command center, or a missile testing range—and then, over the next decade or two, we would arrive by the thousands and tens of thousands until nothing looked or felt as it had before us. Yet whenever we sent our advance teams to some place like the Valley of Heart's Delight, we did not cause panic in the populace; more likely, a flurry of joyous meetings of the Chambers of Commerce and Rotary Clubs. You can understand, then, why families like mine tended to behave with a certain hubris, why in the Valley of Heart's Delight, for example, we were little concerned with a rural society extending back through Spanish missions to acorn gathering Ohlone Indians. We were drawn to the promise of a blank page inviting *our* design upon it. We were perfectly capable of devising our own traditions from scratch if need be.

My mother and father, for example, invented for us certain rites of spring. In the spring of 1962, the Valley of Heart's Delight was covered with blossoms. Back then, the cherry and plum and apricot trees would froth so white and pink that driving around the place felt like burrowing through cotton candy. Spanish colonizers had planted the first of these glades. By the middle of the nineteenth century the valley was a center for growing the "fancy" fruits that needed rich soil, gentle rains, and frostless springs, a Mediterranean soft touch. Just two dozen years before my family's arrival, this was a place of 100,000 acres of orchards, 8,000 acres of vegetable crops, 200 food processing plants, a small city of 50,000, and a half dozen villages that were, as one county planner fondly remembered, "enclaves in a vast matrix of green."

"It was beautiful, it was a wholesome place to live," by that planner's recollection. And every year there would come a day in spring that called forth the blossoms, that seemed to make the world white and pink again in a decisive instant. That was a day eagerly looked toward, no doubt, by the people who had done the planting, the orchard people there long before us.

On warm evenings in the spring of 1962, this is what my father and mother would do. After dinner they would place my baby sister in her stroller and the four of us would set out from the too small, used house they were renting in an established subdivision (already half a dozen years old) named Strawberry Park. We would walk six blocks and run out of sidewalk. We would pick up a wide trail cut a foot and a half deep into the adobe ground, a winding roadbed awaiting blacktop. At a certain point we would leave the roadbed and make our way across muddy clay that was crosshatched by tractor treads, riven by pipe trenches. We would marvel at the cast concrete sewer sections lying about, gray, knee-scratching barrels big enough for me to crawl inside. We would breathe in the sap scent of two-by-fours stacked around us, the smell of plans ready to go forward. Finally we would arrive at our destination, a collection of yellow and red ribbons tied to small wooden stakes sprouting in the mud. These markers identified the outline of Lot 242 of Unit 6 of Tract 3113, exactly 14,500 square feet of emptiness that now belonged to us. All around the outline were piles of cherry and plum and apricot trees, their roots ripped from the ground, the spring blossoms still clinging to their tangled-up branches.

My parents had laid claim to this spot in the usual way. They had sat in folding chairs in the garage of a model home while a salesman showed them maps of streets yet to exist, the inked idea of something to be called Clarendon Manor. They had been given a choice of three floor plans, the three floor plans from which all the dwellings of Clarendon Manor were to be fashioned. My parents had selected the 1,650-square-foot, four-bedroom floor plan, the one with the front door in the middle and the garage door on the right side. They had judged the price of that house—$22,000 at low GI Bill rates and no money down—to be a fair value and just within their budget. They had specified that the kitchen tile be yellow, the exterior trim white, the stucco blue.

My parents had been attracted by some of the features they saw in the Clarendon Manor model home. They liked, for example, the short brick wall with lantern that jutted out the side of the garage, creating a kind of courtyard just before the poured concrete stoop. They liked, as well, the sparkles in the living room ceiling, tiny chips of glass embedded in the white flocking that twinkled by lamplight. They liked these modest nods to tradition and romance, though what they liked most was the functionality of the house's design, the way, for example, that the kitchen, dining nook, and family room merged to created an unbroken expanse of linoleum. This was design for maximum efficiency in the flow of family life, an important selling point for my mother and father.

Here was the deal sealer: By rising early and hurrying to the Clarendon Manor sales office on the day it opened, my parents had been first in line and so had managed to secure a prime lot. Lot 242 was one of very few that stretched wide around the bottom end of a cul-de-sac, a choice cut of land more than twice the size of a standard lot. Naturally, the price was higher: For an additional 8,500 square feet of Clarendon Manor soil, said the salesman, my parents would have to pay $200 more, cash up front. They were only too happy to purchase the extra emptiness.

Once the papers had been signed, the rented house in Strawberry Park seemed to my parents all the more constricting and stale, a house not just used but used up. There was nothing to do, however, but to wait for Clarendon Manor to come into existence, nothing to do but make our visits to Lot 242 on warm evenings. Our rite evolved with the season. Early on, my father would go from stake to yellow-ribboned stake, telling us where the kitchen would be, where the front door would go, which windows would be getting the most sun. Later, after the concrete foundation and plywood subflooring were in and the skeletons of walls were up, we would wander through the materializing form of our home, already inhabiting with our imaginations its perfect potentiality.

Our home, like millions of similar tract homes built throughout America at the time, was said to be "ranch style." Its sober horizontality was said to owe itself to an old-fashioned, Out West wisdom about what a house should be. In truth, the design of our house owed more to a Frenchman named Charles Jeanneret, a man who found his optimism in mechanized shapes, even those (or especially those) made for war. Jeanneret, better known as Corbusier, was that prophet of Modernism who famously declared, "A house is a machine for living in." He wrote this four years after the close of the First World War in his *Towards a New Architecture*, a manifesto containing, as well, these lines:

> The War was an insatiable "client," never satisfied, always demanding better. The orders were to succeed at all costs and death followed a mistake remorselessly. We may then affirm that the airplane mobilized invention, intelligence and daring, imagination and cold reason. It is the same spirit that built the Parthenon.

Corbusier's theory was that houses, like airplanes, worked best when constructed according to rational, "universal laws." One of these laws held that any machine, just like nature itself, must evolve toward ever purer forms. This is why the shapes of progress must look more and more like an airplane, must be ever more streamlined. This is why every bit of sentimental bric-a-brac was wasteful drag holding back our flight into a better future. This is why Corbusier hated Victorian decor "stifling with elegancies" and found the "follies of 'Peasant Art' " downright "offensive." Now was the moment to make "an architecture pure, neat, clean and healthy." For, "We have acquired a taste for fresh air and clear daylight." And, "Everything remains to be done!"

Corbusier's hugely influential "Purism" glorified not only the shapes of machines but the assembly line production machines made possible. He would exploit economies of scale. He would make the parts interchangeable. For the rationally minded new technology worker, he would create "towers in the park" surrounded by greenery and laced into freeways. He would design vast high-rises that stacked families in hundreds of identical compartments, give them "open plan" living areas without room dividers, sit them in no-frills furniture that Corbusier preferred to call "equipment."

You can find the bastard progeny of those towers in skylines from Warsaw to Chicago. Housing projects gray and stark, they are today's emblems of beehive alienation, the worst possible place to look for Corbusier's optimism realized. No, to find that, you would do far better to go to where Purism met the American Dream, places where single-family homes were mass assembled from three blueprints and shopped for like cars, places like Clarendon Manor. In such blue sky subdivisions, Corbusier's tower of identical compartments was unpacked and spread out, forming an architecture all the more "pure, neat, clean and healthy." We who dwelt in them were as Corbusier had predicted. The era's new worker, the aerospace worker, did want to live surrounded by greenery and laced into freeways. We had indeed "acquired a taste for fresh air and clear daylight." And what we wanted were $22,000 Parthenons expressive of the same cold reason we saw in the lines of a jet fighter's fuselage.

There was a man in the Valley of Heart's Delight who made it his business to build the purest tract house forms of all. He was named Joseph Eichler. Joe Eichler had been a rather conventionally minded fellow until the day in 1936 when he happened to rent a home designed by Corbusier's fellow Modernist prophet, Frank Lloyd Wright. With its bold spaces (a long, glass-walled living room) and latest technology ("radiant heating" via water pipes in the concrete floors), the home was to Joe Eichler a revelation, and when the cigar-chewing dairy wholesaler eventually decided to get out of butter and into subdivisions, he hired Wright disciples as his architects. They gave him a three-bedroom house with a sleek flat roof, floor-to-ceiling glass along the rear facade, post and beam ceilings, radiant heating, an open plan interior that spoke of free-flowing emptiness, all of this mass producible with a 1949 price tag of $9,000. By 1967, Joseph Eichler would build some 10,000 houses in Northern California, a particularly large concentration of them in areas closest to Lockheed.

From the street, Eichlers resembled lined-up, identical, earth-toned bunkers, their redwood-sided fronts punctured by the merest, if any, glass. You entered the private realm of the bunker through a door, and then—this was an Eichler trademark—suddenly found yourself standing in an open-air atrium. The atrium, an Eichler sales booster from the day it was introduced in 1957, was the *extra* dollop of emptiness you passed through before you met the true front door of the house and all the glassy walls and clean space within. "The Eichler design stunned us," my father remembers of the first one he and my mother explored on one of their house-shopping expeditions. "The low lines, all that glass. We thought it was a marvelous house. It had this California look to it. It was like nothing we'd seen in the Midwest."

Which, indeed, was the genius of the Eichler design, the way it congratulated its owner for fleeing places so encumbering as, say, the Midwest. Midwest weather made flat roofs and atriums impossible. Midwest people were suspicious of houses with no windows on the street and too few walls inside. A Midwest house (like the lives within it) stuck with tradition. No, the Eichler was like nothing you'd ever see in the Midwest. *A whole new adventure for us,* the Eichler said to its owner. *For everyone!* said the many streets lined with Eichlers, streets that ran in concentric circles closing finally around a swimming pool with a clubhouse, for that was the modern shape Eichler gave the entire neighborhoods he built from scratch.

My father and mother did not buy an Eichler home in an Eichler subdivision, a missed opportunity they speak of wistfully to this day. At the time, the Eichler price, though consciously pegged to aerospace salaries, was a few thousand dollars beyond my father's bottom-rung pay. And yet the allure of the Eichler illustrates why we saw so little "ranch" in the house we did move into in the autumn of 1962. Compared to an Eichler, our house was a quieter shout of Corbusier's machine-minded optimism, perhaps. But our house, too, was "open plan," bright and low and streamlined, laid out unsentimentally enough to please any engineer. Most importantly, ours was nothing that could be mistaken for a used house of the past. Ours was a blue sky house, pure and simple.

One spring Saturday my father rounded the corner of the cul-de-sac wearing his brown leather flight jacket, his hands at the controls of a machine with knobby tires taller than me and an enormous claw upraised. By then we had inhabited Lot 242 for nearly a year, the seasons had come around, and now had arrived the time for making a backyard lawn. Having taken the measure of the hard clay that Clarendon Manor was built upon, my father had devised a plan. The first step was to break the clay into clumps and so he had rented this tractor, had driven it across town and right into the yard through a portion of fence he had removed. To the delight of me and other children in the neighborhood who gathered round, he pulled us up one at a time onto his lap, the rattling of the beast passing through my father's blue-jeaned thighs into our own bodies. We rode the machine as it ripped up the earth, and when the frenzy was over we each took turns sitting in the quieted claw.

Next arrived a dump truck full of redwood chips, backing through the same downed section of fence and stopping at the far end of the yard where a knot of us kids stared up in wonder. "Close your eyes!" shouted my father, laughing as the truck bed rose and a sudden wave of sawdust broke over us, leaving us to rub our eyes as we stumbled out of the pile, all of us now laughing, too. The sawdust was mulch to be blended, according to plan, with the now clumped clay. And so my father next spent several hot afternoons hauling wheelbarrow loads of chips about the yard. The next week he showed up with a new machine that churned the ground with lots of blades, its roar drawing people young and old to watch my father, his face fierce, his T-shirt soaked with sweat, tame the rototiller. After the rototilling was done, my father slipped over his shoulders a harness of rope and began dragging, back and forth over every inch of ground, a heavy, nailed together collection of boards, a tool, he explained, to level the land. When that was finished, on the next weekend, my father invited all the spectating children to rejoin his rite of spring; we were given a nickel for every coffee can full of stones we collected, my father inspecting the haul can by can before dumping each one out in a plastic trash barrel.

Now it was time for my father to draw elaborate diagrams on engineer's graph paper, arcs and intersecting lines, the design of a sprinkler system for our lawn-to-be. When he had calculated the optimum configuration, the one requiring the least amount of pipe and providing the most efficient water coverage, he drove to the supply house that had been established by the old families, the store called Orchard Supply, which now boasted an always crowded do-it-yourself lawn department. The next few weekends of spring were taken up with digging trenches and mastering the art of making joints with PVC pipe. On the evening when he got everything to work, there was a short celebration by the family, all of us watching as my father turned knobs and made water spout from the dusty, flayed ground. In days following, now and then, neighbors would drop by to see the performance repeated.

Finally, my father decided, the time had come to open the plastic sacks that had been sitting since the beginning of spring in a dark corner of the garage. He spread, with a machine for spreading, the grass seed and granular fertilizer. He spread on top of that a half-inch more sawdust, as his studying up in *Sunset Magazine* had told him to do. He rolled, with a machine made for rolling, the entire surface of the planted lawn until our backyard looked as flat and uniform as a paved landing strip. He turned on the sprinklers, and we waited for the grass to come up.

Aspects of the plan, as it turned out, had been slightly off. All that sawdust was not what a lawn in clay wanted, apparently.

My father's sprinkler system was too stingy in its water distribution. And bad grasses called "crab" and "Bermuda" were more interested in living in our backyard than was the Kentucky Blue my father had planted. The lawn did not grow in as smoothly and homogeneously green as my father had expected, but the goal was nevertheless well enough in sight to be pursued. The rite my father initiated that year was therefore to be conducted on a smaller scale every spring to follow for many years to come. There would be new fertilizers and better mulches, new diagrams of ever more optimal sprinkler system configurations. My father would be seen fierce and sweating behind some machine made to punish nature into submission; if not tractor or rototiller, a high horsepowered lawn mower with a fearsome blade. Every spring we children were welcome to gather and watch the original struggle re-enacted until we became old enough to take it up ourselves.

The manner in which we went about conquering the Valley of Heart's Delight—my tribe's methods of infiltration and patterns of occupation—provides a picture of how it often went in many other places.

In the years just after World War II, there was in the Valley, amidst all the green and blossoms, a rich, private university that called itself "The Farm." The Farm wanted for itself a big piece of the command economy, wanted to be a center for federally sponsored science and technology work. And so, in 1951, The Farm created within its borders the Stanford Industrial Park, an enterprise that became a model for fifty similar university-affiliated research parks built across America during the next two decades. Stanford Industrial Park was altogether unthreatening to behold, a campus of Purist boxes nestled amidst lawn and nature, a new form for the technological plant, the "ultimate," as one awed reporter saw it, "in landscaping of an industrial area." Lovely Stanford Industrial Park became a magnet for scholars with something to offer the Cold War project, cutting edge work in the fields of electronics, aerospace, computers. The Pentagon and NASA contractors came with them, some, like Lockheed, locating directly within the Park, others—including IBM, Philco-Ford, Sylvania, and, again, Lockheed—establishing major plants nearby in the Valley of Heart's Delight. Indigenous contractors like Hewlett-Packard and the radar maker Varian Associates boomed with the newcomers, prospering from the high technology synergy (rather than any real competition) created by federal spending in the region. When, in 1955, William Shockley's team came West to be near the Park and to refine his transistor, the Pentagon poured all the more money into the region, snapping up the miniaturized electronics for its missiles, planes, and computers. When other brilliant minds at the Park replaced the transistor with the even lighter and tinier integrated circuit, the Pentagon redoubled its largesse. In 1967, for example, the military bought seven out of every ten such circuits made, microchips that happened to go into the beautiful missiles and satellites of Lockheed, which happened to be the largest single employer in the Valley of Heart's Delight. By 1967, in other words, the Valley of Heart's Delight had become a company town, and the company, in the final analysis, was the U.S. Department of Defense.

We, the hundreds of thousands of blue sky tribe members who came to do the work, wanted our freshly made emptinesses and our brand-new subdivisions, and so the orchards would have to be bulldozed, a fact we disguised by laying our groundwork quietly and ingeniously. As early as 1956 in the Valley of Heart's Delight, we had cast the die. In that year, throughout all the Valley's two hundred-square-mile area, green and blossomy to the casual eye, a scant twenty-six square miles was in "urban use"—yet nowhere in the Valley was there a single square mile without some little subdivision, some small outpost of ours awaiting the full invasion.

"The result was that all 200 square miles were in effect held hostage for eventual development." That is what a pair of our enemies, Samuel E. Wood and Alfred Heller saw all too clearly.

They sounded this and many other passionate, impeccably documented alarms for a group called California Tomorrow, as formidable a voice against us as existed at the time. "Slurb" was the mocking term coined by California Tomorrow to describe what my people tended to create in place of orchards. Slurbs, warned Wood and Heller, were the "sloppy, sleazy, slovenly, slipshod semi-cities" where nine out of ten Californians would soon be living if my people could not be contained, if precious farmlands weren't zoned safe from us, if planning for the good of all could not replace greed at the local level. All this our enemies saw in 1961, their prophecies bound into impressive white papers that went to politicians and newspeople all over the state. All this they saw too late, for by 1961 my tribe had on our side collaborators too powerful and quislings too willing.

We had the mighty backing, for example, of the Federal Housing Administration in distant Washington, D.C., an institution created by Franklin Roosevelt in a spirit worthy of Corbusier. The FHA, like my tribe, was not much interested in a used house, particularly one in any inner city. The FHA, as documented by Kenneth T. Jackson in his noted history of suburbanization, *Crabgrass Frontier,* was most interested in seeing fresh emptinesses filled up with brand-new tract homes. The FHA encouraged this by dangling an enormous carrot before the noses of the nation's private lenders and builders. The business of the FHA was the insuring, with U.S. Treasury funds, of bank loans for housing. But if you were someone who refurbished row houses in urban cores, you could expect a frown at the bank, because the FHA was not willing to insure such mortgages. As Jackson showed, the FHA reserved its sweetest carrot, its highest levels of insurance, for the construction of detached single-family residences for entire neighborhoods of white, middle-income people of non-Jewish descent. For those so favored, FHA insurance trimmed interest rates and drastically reduced down payments, making a blue sky home a near risk-free investment for owner and builder alike.

My tribe found a similar collaborator in the Veterans Administration, the crafter of the GI Bill for the sixteen million veterans of World War II and millions more after them. What a VA-insured loan meant to my father was a mortgage even cheaper than the FHA could make it, a deal two points below the bank's rate, and, instead of ten percent down, not a penny. Without that loan, my father remembers, he who was "cash poor" and who made a mere $143 per week would have been able to afford nothing in Clarendon Manor, nothing around Lockheed but "some cracker box."

My tribe enjoyed the favors of Fannie Mae and Ginnie Mae (the Federal National Mortgage Association and the Government National Mortgage Association). Thanks to the two Maes, any bank could ignore less lucrative local needs and invest in mortgages wherever in America tract homes happened to be sprouting. Fannie Mae and Ginnie Mae, writes Jackson, "made possible the easy transfer of savings funds out of the cities of the Northeast and Middle West and toward the new developments of the South and West."

My tribe found our collaborators in government bureaucracies wherever we needed them. We found one, for example, in A. P. Hamann, city manager of San Jose, the largest town in the Valley. He proudly declared, "They say San Jose is going to become another Los Angeles. Believe me, I'm going to do everything in my power to make that come true." We found one in California's Democratic Governor Edmund G. "Pat" Brown, who ordered a flag-waving, statewide celebration on the day, not long after my family arrived, that California's population eclipsed New York's to become the largest of all the states. When we saw those flags, my people knew they waved in thanks for our coming.

We found the quislings we needed wherever there were orchard people and farmers who might have blocked our plans. We found them because enough money can make a quisling of anyone. Within a decade after the coming of aerospace to the Valley of Heart's Delight, our developers were shelling out nearly

$100,000 per acre for any land left that might still be covered with blossoms. The math is simple enough. One of those acres might have yielded $450 worth of cherries or apricots or plums per year at the time, which meant the acre would have had to blossom year after year for two centuries in order to *begin* to match the amount our developer was offering for it immediately. This should give an idea of the quiet force my people exerted whenever we entered a place, power enough to undo a century-old economy and strip the blossoms from a valley once and for all.

My people did sigh at the extinction of those blossoms. We missed them the way you miss any pretty decoration taken down for good. But honestly, we did not mourn their disappearance in any deeply felt way. Certainly we did not feel guilt. The reason we did not is that those blossoms never spoke to us as they did to the orchard families. We had not, after all, come to the Valley of Heart's Delight to join the circular rhythm of nature. The rhythm we sought to join, the rhythm of Corbusier and Eichler and Stanford Industrial Park and Lockheed, was nothing circular as we understood it. Our imagination was linear, proceeding forward and upward, and our lines did not curve back on themselves as did the seasons. We saw promise in the clean possibilities that arose once every blossom had been erased, never to return.

Those orchard people who held their ground longest, either on principle or for a better price, were phantoms to families like mine. Sometimes from the rear window of the family station wagon, as the pink and white forests whizzed by, I'd catch a glimpse, back in the shadows, of weathered wood buildings, the drooping shape of a barn next to the faded bric-a-brac of some old Victorian. These were the hunkered-in homesteads of the people who used to have the Valley of Heart's Delight. Here and there, too, were stretches of old cannery buildings where fruit was packed by other people we never saw. The city's downtown was a zone of musty hotels and dirty-windowed shops, which is why we rarely went there. The restaurants where the old families ate were shy little turn-ins with something sighingly nostalgic out front like a wagon wheel. It's hard to conjure the images now because at the time none of these places showed much interest in attracting the attention of us newcomers.

Instead, the new roadside buildings on the edges of the Valley of Heart's Delight, our edges, were those clearly eager to please my tribe. So eager, in fact, as to be obsequious. Planets twirled above gas stations, rows of sky-aimed girders turned car washes into rocket gantries, the Futurama bowling alley near Clarendon Manor covered its huge sign with neon stars and amoeba letters. We visited the new eateries and were charmed by the orange vinyl booths and crazily slanted glass walls and stamped-steel boomerangs supporting zigzag roofs. We were charmed enough to invite the shapes to come in off the roadside and into our homes; an example was the clock that seemed to hang over everyone's fireplace, the one with the face surrounded by a sunburst of thin rods with balls on the end. Inspired by *Popular Science* drawings of the atom, the motif came to be known as the "atomic swizzle stick."

Modernist minimalists, the most rigid adherents to Corbusier's vision, scoffingly called such forms "Googie" architecture after the garish chain of Googie's restaurants that began to appear in Southern California in the 1950s. They saw in all the color and flash an affront to their Purism. But the point of Googie, invented by advertising, was to catch the eye from a fast moving car. Googie did so with space age iconography, and so like garlands thrown before invaders, Googie made us feel welcome in Cocoa Beach as well as in Long Beach, wherever in America we established our blue sky outposts. What we saw in a Googie was what we saw in an Eichler, a visual language that not only spoke to us, but *about* us.

It was easy enough at the time to believe that someday the

whole world would speak our language. I remember a Saturday morning not too many years after we had moved into our new home. My mother and father read in the newspaper about a new sculpture erected as a symbol of cultural arrival by our fast-growing city. We drove over to see the thing and when we arrived at its base and looked up, we very much liked what we saw: Benjamino Bufano's "The Universal Child," big blue eyes atop a tapered stainless steel cylinder shaped like a beautiful missile.

Several valleys over from ours, Joan Didion watched the coming of my tribe with dread. We moved her to write, in a 1965 essay, how it felt to be a "native daughter," to have "come from a family who has always been in the Sacramento Valley" and to see that "the boom was on and the voice of the aerospace engineer would be heard in the land. VETS NO DOWN! EXECUTIVE LIVING ON LOW FHA!"

Fifteen thousand aerospace workers, "almost all of them imported," had arrived on the outskirts of Sacramento to join Aerojet-General, a maker of missile boosters. Joan Didion's family was, like the orchard people of the Valley of Heart's Delight, a family tied to agriculture with a hundred years of circular rhythms behind them. Hers were a people primly insular and tragic minded, according to the native daughter. Her Valley was a place where "incautious" children visiting from out of town often would drown in the river, disappear forever, and the old locals would see a proper lesson in that, would say, as Joan Didion's grandmother did: "They were from away . . . Their parents had no *business* letting them in the river."

Joan Didion saw fifteen thousand out of towners coming to stay forever and concluded that Sacramento had by 1965 lost its "character," that because of us it was "hard to *find* California now." She looked at the children of Aerojet-General and thought . . .

> Their grandmothers live in Scarsdale and they have never met a great-aunt. "Old" Sacramento will be to them something colorful, something they read about in *Sunset.* They will probably think that the Redevelopment has always been there, that the Embarcadero, down along the river, with its amusing places to shop and its picturesque fire houses turned into bars, has about it the true flavor of the way it was. There will be no reason for them to know that in homelier days it was called Front Street (the town was not, after all, settled by the Spanish) and was a place of derelicts and missionaries and itinerant pickers in town for a Saturday night drunk . . . They will have lost a real past and gained a manufactured one.

In another essay written five years later, Didion gets at the profound difference between her people and mine. She writes of "growing up convinced that the heart of darkness lies not in some error of social organization but in man's own blood." She reveals herself, in other words, to be a pessimist about human endeavor engineered and executed on a grand scale. How different from my tribe, who would say instead: If incautious children might drown in a river, let us erect a Cyclone fence, even drive the river underground, leaving behind a manufactured surface that was dry and safe, empty and speaking of promise. That, after all, is what was done with the creek that ran by Clarendon Manor.

A past *manufactured,* Didion lamented in 1965, feeling sorry for any boy growing up like I was. But I remember a great hulk of unmanufactured past that sat by the freeway a short drive from Clarendon Manor, the mansion of Sarah Winchester, wife of the rifle tycoon. The Winchester Mystery House (as it was called on billboards with skulls leering) was begun in 1884, and old Mrs. Winchester never stopped construction until the day she died in 1922. In recent times, the house and grounds have been renovated (or "remanufactured," as Joan Didion might say), but when I was a boy it was all crumbling authenticity. I would sometimes be taken there for a class field trip or just something to do on a rainy day. A guide would lead us through rooms that had never been used, show us staircases that led to nowhere, point out windowpanes and tiles and even sink drains that, by Mrs. Winchester's superstitious orders, always added up to thirteen. She believed she was haunted by wicked spirits angry over the killing her husband's rifle had done, haunted, too, by good spirits who would ward off the bad ones as long as work continued on her house. She was as crazy as she was rich, and so her house with its one hundred and sixty rooms and forty-seven fireplaces and thirteen bathrooms stretched across six acres, a horror house to me simply for its dirtiness and darkness and wrongheadedness.

I did not like visiting the Winchester Mystery House, did not think much of Mrs. Winchester and her unvarnished past. I thought she was a silly old lady to cower in her morbid pile, as silly as were her bygone times. She was a wild extrapolation of what I imagined the orchard people to be, the best reason for there to be streamlined tract homes in clean subdivisions. We children would take the tour, and the dust and strangeness would tire us out and we'd be happy to emerge into the bright daylight. Happy to be free of the stifling *obsolescence* of the place.

If you had pointed out to any blue sky invader on the Mystery House tour that the new money in the Valley of Heart's Delight was thanks to weapons potentially vastly more destructive than all the Winchester rifles ever made, we would have shrugged and laughed all the more at Mrs. Winchester's guilt-driven insanity. We were not much interested in ironic abstractions. Rather than visit the Winchester Mystery House, we far more often visited the places directly next door, the Century 21, Century 22, and Century 23 Cinemas set back on a broad plain of parking lot asphalt, movie houses that were low and round and shallow-domed like flying saucers. The whole family would set off to see the movie that everyone was seeing, the big movie called *The Sound of Music*, and we would park our car between a Plymouth Satellite and a Dodge Polaris, and we would join our fellow citizens within the glowing belly of one of the spaceship cinemas, and there on the screen would be a Catholic nun, full of fun, singing to children with our own ruddy cheeks.

We were a tribe with hubris, as I say, but you can see how we came by it. By the time we were done with a place, everything around us seemed to cheer us on. Everywhere there were signs telling us that by moving to this empty new corner of America, we had moved closer to the center of the nation's imagination.

Joan Didion did not like it that such a tribe had found its way to the outskirts of her Sacramento. Now that the people of Aerojet-General were using the latest materials to manufacture optimism on the same farmland that had made her people hard and tragic, Didion was forced to rethink what might be the "true" California. "Which is the true California? That is what we all wonder," the native daughter wrote in 1965. Was it the dusty Main Streets of the Central Valley that she knew so well from her childhood? Or was the "true" California her vision of my tribe, "the legions of aerospace engineers who talk their peculiar condescending language and tend their dichondra and plan to stay in the promised land"?

If she had asked my father the aerospace engineer or my mother who had joined him on a whole new adventure, they would have told her: There is no "true" California, only latest improvements in design. Our fellow tribe members in Seattle or Houston or Colorado Springs or Bethpage or Cape Kennedy or Huntsville would have told her the same thing about the "true" Washington, the "true" Texas, the "true" Colorado, the "true" New York, the "true" Florida, the "true" Alabama.

Within twenty years of the opening of Stanford Industrial Park, within ten years of my family's arrival, the Valley of Heart's Delight was no longer a place of "enclaves in a vast matrix of green." It had become a vast matrix of expressways and freeways and Clarendon Manors, a vast matrix of companies making technology primarily for the government. The population had grown many times over in those two decades, and we no longer heard the Valley of Heart's Delight called that anymore. In fact, no one I knew had ever used that sentimental name. While I was growing up, my family simply had called it the Valley, or, as it was officially termed on the government studies and the plans of various developers, the Santa Clara Valley. It would not be until a time distant, well into the 1970s, that we would begin calling our home Silicon Valley.

There was yet another rite of spring practiced by my family, a rite that became possible once the occupation was all but complete, once nearly all the blossoms had been replaced by settlements like our own. On an evening that was bright and windy but too warm to be winter anymore, my father would come home from Lockheed with a kite or two, balsa sticks wrapped tightly with colorful tissue paper. If the next morning was a Saturday, he would put the kites together for us, tear us a tail from an old sheet, make a string bridle that held the kite just so, help us launch the kite and send it up over the tract homes. For just this very purpose, my father kept what seemed a mile of twine on an enormous spool, and so the kite would climb higher and higher until it became a shimmying dot against the blue.

At that point my father would go into his garage and make a small parachute. He would unfold a paper napkin and tie its corners to four strands of string, drawing the other ends of the string together and knotting them around a bolt for weight. He would stick a bit of reinforcing tape in the center of the napkin and pass through that a bent pin, making a hook that poked out of the top of the parachute. Next my father would write our phone number on the parachute with the words: IF FOUND, PLEASE CALL.

"Ready for takeoff?" my father would say as he grabbed hold of the taut kite string and hooked the parachute onto it. And then a miraculous thing would happen. Driven by the wind, the parachute would skitter up the line, joining the kite high in the sky in what seemed an instant. When it reached the top my father would say, "Give 'er a jerk!" and the parachute would fall away from the kite and drift in whichever direction the wind was blowing until we could see it no longer.

Then would begin the wait for the phone to ring, the wait for someone to call and say they had our parachute. If hours went by, my mother might suggest a prayer to St. Anthony.

Tony, Tony, listen, listen.
Hurry, hurry, something's missin'.

"You have to believe."

If we said the prayer and did believe, the ring would come and someone would say, "Got your, uh, I guess it's a parachute, here. Landed in my backyard. Almost ran over it with the mower." My father would write down the address and he would get out the street map. He would pinpoint our destination, and we older kids would set off on our Stingray bikes, having been given a reason to trace a route we never would have traced otherwise, so empty and so much the same was every street for miles around. We would leave our cul-de-sac named Pine Hill Court (where there was neither a pine nor a hill) and we would pedal far beyond Springwood Drive and past Happy Valley Way, ending up in some cul-de-sac we had not known existed. And there would be a man about my father's age with a similarly receding hairline and knit sport shirt, a man who seemed to be pleased at the serendipitous fun our parachute had brought into his Saturday, a man who safely could be assumed to do blue sky work for a living. Anywhere we cared to drop a parachute from the sky there would be someone like him, a house and family like his and ours. That is why I think of our game as a spring rite for blithe conquerors.

THREE

OUR LADY OF AEROSPACE

Where is God? asks the Church.

God is everywhere, the child of the Church is taught to answer without hesitation.

The Catholic God is everywhere, my mother was told again and again by the nuns and the priests, by her Irish Catholic father, by her Luxembourguese Catholic mother, by two sisters her elder by many years. Growing up in Rock Island, Illinois during the 1930s and the Second World War, a girl found it natural enough to believe that the Catholic God fully inhabited such a place. Life in that town by the Mississippi was nothing sure or easy, just as the Church, in its sad wisdom, had preached for two millennia. For most of her girlhood, my mother slept on a pull-out couch in the dining room of a pinchingly small house, the best her father could provide on his pay as a cost estimator for the local arsenal. There had been a time when he had been boss of a company that made bolts for the railroads, and they all had lived in a roomy, even fashionable, home in New Jersey. But the Depression had taken that away, and all my mother remembers is her father as he was in Rock Island, a man who was kind and patient and God-fearing after many worldly disappointments.

Rock Island's churches were dark, full of flickering votive candles that lit the patient faces of the saints and the Blessed Mother. The steaming summers gave a hint of what suffering awaited a selfish sinner. The winters were as pitiless as the Sunday sermons. "What we heard a lot from the priests back then," my mother remembers, "is that we'd better shape up or we were all going to go to hell." Rock Island was the sort of Midwest smudge of industry and docks that daily made its people not expect too much, daily required hope and faith. *Where is God? God is everywhere. Even here. Especially here.*

Yes, but could the same Catholic God be found if one lived in prefabricated paradise? Would the same Catholic God be there?

That was a popular question in 1960, the year my mother set about answering it, the year we moved to the Valley of Heart's Delight. There were, for example, the intellectuals who assumed God dead, murdered by scientific rationalism. They wondered if anyone as modern as an aerospace family could go on believing in Him. And, too, there was the Catholic leadership, whose American Church had thrived in the cities' ethnic enclaves. They nervously watched to see if suburban soil would prove as fertile. All of them were very interested in my mother, the Catholic daughter of Rock Island who had gone and married a Protestant Lockheed engineer and now was moving into a brand-new subdivision in Northern California. Could her God withstand the pleasant weather, the booming economy, the faith in progress, the willed rootlessness that drew a generation like her to such places?

My mother, for one, did not see why not.

She did not see why her children could not be raised to be believing Catholics in a space-age cul-de-sac, could not be taught to say without hesitation: *God is everywhere. Even here. Especially here.*

The Monsignor Irvin A. DeBlanc, who was at the time the director of the Family Life Bureau of the National Catholic Welfare Conference, had a word for the threat posed by my mother's marriage to a Methodist. "Leakage," was his term for what too often came of such unions, the dilution, the drying up of the Catholic way of life. Catholics marrying non-Catholics represented an outflow "much greater than the adult converts we are making every year," Monsignor DeBlanc warned attenders of the Catholic Family Life Convention in San Antonio, Texas. His words found their way to our blue sky suburb via *The Monitor*, the official newspaper of the Archdiocese of San Francisco, sold in foyers of churches throughout the region. On July 1, 1960, the day Monsignor DeBlanc's thoughts on leakage appeared, *The Monitor* gave them its boldest headline: "Expert Sees U.S. 'A Generation from Paganism.' "

If my mother saw that headline staring from a newspaper rack, she would have been in the foyer not of a church but of a gymnasium, for that is where she was attending Masses at the time, in the gym of Mother Butler Catholic High School for Girls in San Jose, California. The pews were folding chairs, the altar a linen-covered cafeteria table. The congregation of hundreds was made up, for the most part, of people like my mother, young Catholics following the future West. The local Mass was this one because a Catholic church to serve the area was still on the drawing boards, its intended site a plum orchard.

As soon as she and my father (with me three years old in the back seat) pulled into town in the green, tail-finned Ford that brought us from Springdale, Ohio, before the pictures of the Grand Canyon taken along the way were developed and even before the moving van showed up in the driveway of our rented ranch house, my mother asked neighbors where a Catholic Mass could be found. She prayed in the gymnasium, that first Sunday, with me on her lap and her Protestant husband by her side.

This was in keeping with the understanding my mother had reached with my father as a condition of their marriage. The children would be raised Catholic; my mother, therefore, would head up the spiritual affairs of the household. My father, while unwilling to convert to Catholicism, otherwise had no quarrel. Already, he had let his Methodist inheritance lapse, preferring to place his faith in the provable methods that made jets fly and satellites orbit. "I don't go to any church, anyway," my mother remembers him saying, "so what difference does it make?" My mother saw in his words nothing to suggest the threat of leakage. She believed she had sealed an agreement to bring into this world more Catholics. She intended to introduce her children, however many there might be, to the Catholic God she worshipped.

This God was not the same as the scowlingly gothic God whose priests threatened hell to a working-class girl in Rock Island. That gothic God had not been much interested in worldly striving toward modernity. *Contemptus mundi* was a phrase He had enjoyed hearing his supplicants say since the Middle Ages. In Rock Island's old Catholic churches, as it had been forever, He spoke through a hierarchy of chosen men whom He dressed in lace and gold filigree, priests who whispered the Mass with their faces to the wall while those in the pews looked on, silently mindful of their place at the bottom of the eternal chain of command.

My mother learned as a girl all the many rules laid down by the gothic God, the categories of sins to confess, the fastings at Lent and before every Holy Communion, the fact that a bad thought is recorded in heaven just as surely as an evil deed. My mother came to know the age-old Catholic lesson of complicity, the inescapableness of guilt. Guilt was the voice that said, "If you do this thing for yourself, you are not acting in service to others. You are selfish." Guilt was the reminder that every selfish act sets off a ripple of pain that connects somehow to all the misery on the planet, a ripple that reaches and pains and angers God Himself. Guilt, as wielded by a gothic God with contempt for this world, might well have prevented my mother from imagining herself raising a Catholic family with a Protestant husband on the California frontier.

She preferred, instead, to see her California life as an *answer* to her prayers, a blessing bestowed by a God shedding his gothic temperament in favor of the optimism of the times. It was my grandmother who provided her the first glimpse of this blue sky God. My grandmother, says my mother, differed from the official voice of the priests in that she was "a lighthearted Catholic," the kind of Catholic who would burst out giggling in the middle of saying the Rosary with neighbors, who found it easier to give her faith to a merciful God with a sense of humor. My grandmother was a school teacher who spent many evenings tutoring an illiterate Italian couple down the street and many more evenings guiding her husband through the correspondence courses that finally earned him a high school diploma when a middle-aged father of three. In action and prayer, my grandmother made my mother imagine a God with a loving eagerness to see His children enjoy success in this world. This God, when my mother was eighteen, resided in a place called Marycrest, the new women's college that stood on a bluff across the Mississippi overlooking Rock Island's old churches. My mother was given a scholarship to attend Marycrest and learn from the Sisters of Humility the science of biology.

"We were serious," says my mother about her best friends among the several hundred Marycrest students. "We thought we could do anything. And we were anxious to leave Rock Island for some place warmer." She did not feel guilty about such aspirations, did not consider them selfishness, because the Sisters of Humility would not let her. They were "brilliant women" who believed in the freedom that education afforded women, a freedom that allowed my mother to graduate with her biology degree and use it to land a hospital job in Corpus Christi where there were parties around swimming pools thrown by naval aviators with exciting prospects.

In the years just after moving to California, remembers my mother, she read *The Monitor* faithfully and for a time subscribed. ("A MUST in every Catholic home," was the paper's slogan.) Wanting a sense of what my mother would have heard the Church saying to her, I have gone to archived editions of *The Monitor* from those times. I have found on some pages the voice of the gothic God, the voice of a Monsignor DeBlanc. But far more often I find a Church advancing its claim on blue sky optimism.

Headline for June 17, 1960: "Catholic 'Role' Grows." A sociologist at the University of Utah, reports *The Monitor,* finds that Catholics "are moving up into" what he calls "the 'upper middle class,' progress made possible because college training has replaced money 'as the chief avenue of social mobility,' because ethnic identification is becoming relatively unimportant and 'spiritual ghettos' are coming to an end." That is why the "next 30 years will see Catholics taking larger and more important roles in U.S. life."

Headline for July 29, 1960: "Vast Building Program Within Archdiocese." The expansion of the Catholic infrastructure, "new churches, schools, convents, rectories and hospitals" makes for a list of names (St. This and Our Lady of That) taking up thirty column inches. My mother is in the company of so many fellow Catholics in this, one of the nation's fastest growing regions, that church construction is of "staggering proportions."

Headline for September 23, 1960: "Church Given Higher Rating For Efficiency." Out of an optimal efficiency rating of 10,000 points, the American Institute of Management has awarded the Catholic Church 9010, up from five years before. The latest pope, John XXIII, is judged to have brought " 'a completely new spirit' to the management of the Church," combining strong leadership with an up-to-date belief in "the principles of delegation and decentralization. . . . All down the line there has been a noticeable improvement in placing the right man in the right position of authority."

Headline for August 19, 1960: "Polaris." Beneath runs the opinion of a *Monitor* columnist who endorses the work of men like the one my mother has married, makers of blue sky weapons. "The story of such a weapon—the Navy's Polaris missile—will be told this Sunday on 'Our Catholic Heritage' at 10:30 A.M. over KGO-TV (Channel 7). . . . This column has pointed out before that a steady diet of 'Red superiority' propaganda can sap the American will to resist. . . . The world knows now—and especially the Communist world—that the Polaris weapons system is an unclenched fist about the throat of a would be aggressor. One move by him and the fist contracts instantly."

Appearing in the July 1, 1960, edition, running in the left-hand column hard against the headline "Expert Sees U.S. 'A Generation from Paganism,' " there is this other headline: "Democrats Will Pick Their Man." Below are pictures of several well-known politicians, including John F. Kennedy, the handsome, young, and energetic Irish Catholic who will seek the presidential nomination later in the month.

In the May 11, 1962, edition, I find what I find in the back pages of most editions: pictures of new schools and churches rising in the suburbs. One of the pictures this time is of the just completed Queen of Apostles, the church that will be my mother's and her family's. The architecture is aggressively Modernist. Temples of all faiths being built in the region at the time tended to be aggressively Modernist, constructed, Corbusier might have said, as machines for worshipping in. Queen of Apostles is unpainted cinder blocks and red roof tiles, a stripped-down echo of Spanish mission with a cross-topped needle rising high above it. Inside, beams of laminated wood sweep up to support a wide, open ceiling of knotty pine. Behind the altar rises a rock wall and a huge crucified Christ. There are no stained-glass windows, no banks of votive candles, no grottoes with statues. The architects, according to *The Monitor*, say the genre of Queen of Apostles is "modern gothic."

When the Valley's ongoing aerospace invasion made another Catholic church necessary not many years later, the "gothic" was dropped altogether. The church erected a few miles from Queen of Apostles was round and filled with sunlight and abstract religious sculpture. It looked like a flying saucer and it was named Church of the Ascension.

Not long after Queen of Apostles opened its doors, my mother called its pastor, the Reverend Elwood Geary, to say, "Father, I can see the top of your steeple from my dining nook window." Not long after that, Father Geary arrived at our doorstep to bless our tract home with sprinkles of holy water. A few years later, Queen of Apostles Grammar School opened for business with just over a hundred students, grades one through three, the other grades to be added, one per year, until there were eight. My mother made sure that I was among the first enrolled, a second grader in red sweater and clip-on tie and polished leather shoes, learning from nuns in black habits all the prayers my mother had learned as a child.

"O my God I am heartily sorry . . ." I would mumble as I paced the orange shag carpet in our living room, groping through the Act of Contrition the nuns had assigned me to memorize.

"For . . . ?" Dad would prod from the couch, scanning the dittoed blue text, just as foreign to him, that I'd brought home.

"O my God I am heartily sorry for, uh, uhhhh . . . uh, oh."

"For having offended Thee. From the top . . ."

Only when my father was satisfied that I would not embarrass the two of us was it time for me to recite the prayer to my mother, who would nod encouragingly as she ironed shirts or folded diapers or snapped Tupperware lids over leftovers. The next week it would be the Act of Faith or perhaps the Apostles' Creed.

My mother assumed the task of making us not merely Catholic, but Irish Catholic. Although at least half a dozen other European bloods mixed within us, she decided her children would be Irish like her dead father because, I think, all of the up sides to Irishness appealed to her. In inventing an ethnicity for us, she selected only Irish positives, giving us to understand that we were genetically impish and fun-loving, not unlike the leprechauns who lurked in the oleanders she'd just planted along our backyard fence. She made Patrick my middle name, although everyone in the family, by dint of our willed Irishness, was supposedly on special terms with the good St. Pat. Somewhere deep in her closet my mother kept a stash of green shamrocks and *Kiss Me! I'm Irish!* buttons that came out on St. Patrick's Day; no child left for school without one pinned to his or her sweater. Like our Catholicness, this was an Irishness free from the scold of the Old World. Ethnicity need not bind one to outmoded tradition. Ethnicity, in this place scrubbed of cultures, was a flexible tool for the enhancement of personal identity.

The great advantage to moving into a freshly made emptiness is that all your favorite beliefs can be unpacked and arranged anew, everything else left behind or stowed away in a cardboard box rarely, if ever, to be opened again. As soon as we moved in, my mother chose to lay a patina of low-key mysticism onto the empty surfaces: a small plastic statue of Mother Mary over the kitchen sink, the Holy Family huddled atop the stereo console, Christ on His Cross on every bedroom wall. My mother chose to bring her children to Queen of Apostles on the Feast of St. Blaise so that we might kneel before the priest, crossed candles in his hands, our chins lifted and the waxy coolness of those candles pressed against our throats while the priest asked Blaise the Throat Healer to protect us from tonsillitis. She insisted we be at Mass every Sunday and Holy Day of Obligation, that we have our foreheads smudged with charcoal on Ash Wednesday, that we give up candy for Lent. At my mother's urging I became an altar boy, and she was the one to wake me before sunrise many mornings so that I could be the only soul riding his bicycle through the subdivision streets, hurrying to arrive at church in time to serve the 6:30 Mass. Answering my knock on the door of the sacristy, a priest would let me in, nod to me, and return to the slipping on of his vestments while I, in the next room, buttoned my black cassock. Those were days when a Catholic mother could imagine her son alone with a priest in the sacristy before dawn and feel certain she had placed a young boy in the safest place in all the world.

On Fridays the practice air raid siren would sound and the nuns of Queen of Apostles school would send us under our desks, having given us scapular medals to wear around our necks that would—provided we remembered to say a Hail Mary just before the sky exploded—secure us immediate salvation in God's heaven beyond my dad's heavens.

That a prayer to the Virgin Mother would hold final sway with God the Father made intuitive sense to me. In our home, where religion was clearly my mother's realm, godliness was, ultimately, feminine. The manner of the priests of Queen of Apostles did nothing to counter the impression. If they were not feminine, they were men wholly unlike my father. Godliness was soft like their hands and voices; the making of a miracle asked nothing more physical than the raising of the chalice above the head. They were quiet men with quiet eyes who were more often seen wearing gowns than trousers and none of them projected the muscular presence of my father, a man who was quick to shout and to perspire, who would come home from Mass and change into jeans and T-shirt and grab tools and crawl down into the dirty spaces underneath the house or the cars or the bathroom sink.

How neatly and seemingly naturally fell the divide between the domains of father and mother in our home. My mother, the repository of mystical information; my father, with a scientific explanation for everything this side of heaven. My mother, who spoke in elliptical, musing patterns; my father, who started at "a" (often literally) and proceeded on to "b" and "c" and "d" until his point could not be mistaken. My kitchen-based mother, whose charge was the seamless maintenance of household operations, all vacuuming, scrubbing, laundering, grocery shopping, and preparation of meals; my garage-based father, who was the wielder of tools against sudden crises, the dead vacuum cleaner, dryer, car, TV. My mother, who on a bad day imagined herself a suffering martyr, her Catholic guilt preventing her from demanding help from her husband and children until guilt gave way to self-pity and her tears of frustration flowed; my father, who on a bad day was not a bit hesitant to let go his temper and demand at the top of his lungs. My mother, who on a good day made us a lemon meringue pie from a recipe in *Sunset* magazine; my father, who on a good day made us patio furniture from plans in *Sunset* magazine.

The naturalness of this divide appeared evident to me in the two great holidays of blue sky suburbia: the Fourth of July and Christmas. The Fourth was a celebration of my father's realm. My mother played a respectfully supporting role, joining with the other mothers to prepare the picnic food that everyone would eat at tables set up in the middle of the cul-de-sac. We children ate with one eye on the fathers, watching for the moment when they would gather and place twenty-dollar bills in the hands of one of their members (my father, usually), who would set off for the fireworks stand. The culmination of the Fourth always arrived in our backyard under the direction of my father. While the gathered neighbors clapped and cheered, he would apply the flaming tip of his blowtorch to the Volcano Cones and Pinwheels and Piccolo Petes he had arranged on a board spanning two sawhorses. The Fourth, a civil holiday, a masculine holiday, was a day for a father to demonstrate his affinity for fire and smoke and noise, the elements that made jets and rockets go.

If the Fourth was loud action come and gone in a flash, Christmas was a gradually building season of mystical imagination. My father lent a man's sweaty energy where needed—the cutting down of a tree at a mountain farm, the stringing of lights along the eaves of the house—but my mother, she and her Catholicism, provided the spurs to the imagination. I knew that the plastic figurines of our manger scene, so old that the paint was wearing off, had been hers forever. She was the one who sewed our costumes for the school Nativity pageant. Under her direction we prayed over an Advent wreath of pine boughs and candles at dinner every night for four weeks leading up to the big day. When the day came, after the presents were torn open, Christmas would culminate with the feminine, the spiritual, everyone saying my mother's prayers at Queen of Apostles mass.

Were these realms, my father's and my mother's, so separate as to be the enemy of each other? On the contrary, the marvel to me is how comfortably they seemed to co-exist in my child's mind. There is a Christmas morning snapshot of me at the age of six or seven, my thrilled face looking up at the camera as I show off my favorite present that day, a big toy aircraft carrier complete with fighter jets that catapulted off the deck. My father the former Navy jet pilot no doubt wanted me to have this "Mighty Mathilda" as much as I did. In the picture a presence is visible behind me, a presence created by my mother using modern and practical materials, Styrofoam backing and purple plastic foil and a tinsel-wrapped pipe cleaner for a halo. The angel my mother has assembled and mounted over the family room fireplace is smiling down on my Christmas morning.

A modern Catholic did not grasp for status. A modern Catholic welcomed the uniformity of Clarendon Manor's houses, which allowed us to imagine ourselves free of the trappings of class. We Catholic schoolchildren flowed out of our modestly alike homes wearing modestly alike uniforms. We gathered in the school yard, formed rows of red sweaters, recited in unison the Pledge of Allegiance, sang in unison "The Star Spangled Banner" before filing inside for prayers.

Somewhere far away in the world there were Catholics

much poorer than us, Catholics with brown skin slow to join the future. I knew this because I had seen their faces on the charity envelopes at Queen of Apostles Mass. It did not occur to me (until long after they were gone with the orchards) that the men and women and children climbing ladders into the fruit trees around our subdivision were Catholics. Their brown faces never appeared in the pews at Queen of Apostles Mass, nor the desks at Queen of Apostles school, so how would I have known? Almost all the children at the new nearby public school were white as well, as was the face in every kitchen window, behind every lawn mower. Parishioners who lived close by Queen of Apostles liked to refer to their neighborhood as "the Catholic ghetto," as if to chuckle at the propitious accident that placed people with so much in common in the same place.

But it was no accident, our happy homogeneity. An original sin lay at the root of it. My mother remembers what the salesman for Clarendon Manor told her and my father the day they signed the deed for Lot 242. "Your investment in this home is a safe one. We aren't selling to colored people." That the salesmen in other model homes throughout the area were making the same assurances was an open secret back then, my parents remember. The builder Joseph Eichler, brave in his ethics as well as his tract home architecture, was virtually alone among his local peers in denouncing the practice. As for the poor, they were forced beyond the boundaries of Queen of Apostles by more subtle techniques, zoning laws and red-lined lending, invisible barriers to the existence of any low-income housing amidst the blue sky settlements.

During recess at Queen of Apostles school, we freckled ten-year-olds would shout "nigger!" at each other with delight. Whether some of my friends were repeating a word they heard mothers and fathers say at home, I can't say. I think I know why I joined in. I shouted "nigger" because there were no blacks to hear me. And because I lived in a place where there were no blacks to hear me, "nigger" was the most exotically evil word I knew. I sensed that the word was a supreme violation of what modern Catholics in a classless subdivision were supposed to feel, the tolerance and even special (if abstract) warmth for people of other races. I knew that my parents were in full agreement with the priests and nuns of Queen of Apostles that black people and brown people, wherever they might be, were really just like us, and that they deserved to live like us and that probably, someday, they would. I don't think my mother was aware at the time that Cesar Chavez had been, some years before our arrival, one of those who climbed ladders in the orchards very near where Queen of Apostles now stood. But she knew that he was a symbol of the modern direction the Church was taking in its relationship with the poor and the brown-skinned in the hinterlands where those people were to be found. She knew that Chavez and his marching farmworkers sang Catholic hymns and held aloft images of Our Lady of Guadalupe, and that *The Monitor* approvingly reported on those faraway struggles, indicating to its readers that social justice was the business of all forward-thinking Catholics.

She knew this and made me know this, and so one day I was invited to step before my white schoolmates and receive a prize for the message of modern tolerance and good will I had crafted in art class with crayons and butcher paper. My poster carried a slogan I'd made up myself: *People Come in Different Colors*. I hoped for a special reward. When particularly moved, the nuns were known to bestow a bit of cloth or bone they assured was the relic of an ancient saint. This time the plastic case I received contained a relic with a different sacredness, a scrap of burnt foil from a *Gemini* capsule, skin that was singed, I was told, as it fell back to Earth.

The year in which we moved into our new tract home and Queen of Apostles rose into view from the window of our dining nook was the year in which Pope John XXIII convened his Second Vatican Council in the name of *aggiornamento*—the bringing of the Church up-to-date, the modernization, no less, of the Catholic understanding of God's will.

Now the American Mass would be said in English, the priest wearing humble vestments as he faced a congregation who spoke back loudly and confidently. Now church interiors would be democratized, the altar moved off the back wall and railings around it taken down. Now the gothic God's trembling children would be treated as capable adults, the spiritual and practical workings of the parish placed more directly into their lay hands. Having been brought up-to-date, the Catholic God would encourage the popular taste for peppy, catchy folk hymns. The Pope of this God, meanwhile, would sound quite a bit like my grandmother, proclaiming the Divine preference for a "medicine of mercy rather than severity." Pope John XXIII would say this as he called for a new day of Christian unity, an "ecumenism" that recognized that the Catholic and Protestant Gods were in fact the same One with different faces.

Every one of these modernizations my mother embraced fully and eagerly, as if the Pope and his Council had read her newly suburban, religious mind. As Catholic historian Jay P. Dolan writes, Pope John XXIII

> . . . cut through tradition like a hot knife passing through butter, simply but decisively. Gone was the tiara, seclusion in the Vatican palace, aloofness, the trappings of imperial splendor, and harshness towards people of other religions. In their place stood warmth, concern, openness, simplicity—an urbane, modern style not unlike that of John F. Kennedy.

Kennedy, yes. For my mother and thousands like her, the family in the White House was that other, irrefutable sign that heaven smiled on the forward-thinking. There was, for example, the photograph appearing everywhere at the time: Jackie Kennedy in her sleeveless linen dress, her lace kerchief tied lightly under her chin, and her white gloved hand guiding John John to church on Easter Sunday. This very well could have been a picture of my mother (another slender brunette, smiling shyly) and me. This picture was, then, the nation's latest image of grace: Catholic Jackie, modern Jackie, whose modern Catholic husband was vowing to send a man to the moon. The God of the Kennedys was my family's God, an incarnation of the God worshipped, in a kind of ecumenism, by our entire tribe. He was a God who unreservedly endorsed progress, personal and national, and so the new suburban way of life that went with it.

What did this God ask of his chosen people once they had arrived in the promised land? Less than He asked in Depression-era Rock Island, perhaps. But also more. In the third grade, I listened to Sister Elizabeth tell of a boy in another land and time, poor and about my age, who was sweeping out the cobwebbed attic of an ancient church when he discovered a life-sized crucifix. The boy, Sister said, prayed with such belief that Jesus suddenly stepped down off that Cross, touched the boy, and gave him his blessing. I was excited to know that such things could happen, then sad in my gradual conviction it could never happen to me. Our church had no forgotten corners where anything old might lie undiscovered. There were no bells in its bell tower, just loudspeakers that played recorded chimes. Queen of Apostles was an environment of stackable steel chairs and checkerboard linoleum, and Jesus did not appear in places smelling of fresh spackle.

"A View of the Church in America's Growing Suburbia" promises the headline in the March 11, 1960, edition of *The Monitor,* in which a San Francisco priest offers his review of *The Church in the Suburbs,* a book by the Reverend Andrew Greeley. "In thousands of packaged communities across the land, a new way of life is taking hold, characterized by middle-brow conformity, groveling status seeking, and jolly togetherness," summarizes the Jesuit reader. "But there are cracks in this picture window society [that] spell more trouble than crab grass in the rectory lawn." Father Greeley is described as an up-and-coming sociologist who agreed with secular works of the day like David

Riesman's *The Lonely Crowd* and William H. Whyte, Jr.'s *The Organization Man.* Such experts, whether they wore Roman collars or not, worried that too many families like mine were turning into emasculated corporate drones and mad housewives and spoiled children. The new suburbia, they warned, would test our souls with its soullessness.

My mother did not read such books, did not begin to comprehend such pessimism. In the synthetic shine of a new church, in the tape-recorded chimes of Queen of Apostles, she saw and heard what her God expected of a modern Catholic mother. Some evenings she and I would walk the several blocks to Queen of Apostles and look for the green light over the confessional that signaled a priest was ready to hear our sins. I would enter the closet next to the priest's and I would marvel at how the whirring fan and lamp inside were rigged to turn off as soon as my knees touched the kneeler. I would be glad for how efficiently the Catholic process in this modern world flowed, the sliding back of Father's window, the tallying of my sins for Father, his doling out of a penance that was reliably ten minutes' worth of Hail Marys and Our Fathers. I would briskly give myself over to the receiving and achieving of short-term goals that was this process, rattling off the prayers as soon as I'd left the confessional. And then my mother, who was sitting in the next pew, would be done saying her own penance and her face would come out of her hands and she would smile at me. She would cause me, then, to connect the utilitarian with the eternal. She would remind me that the evening's process was for the care and maintenance of the soul, the soul that was *(even here, especially here)* very real, real enough to be not just imagined but *felt.* She would do this with her usual cheerful air, saying to me as we left our church, "Now. Don't you feel better?"

FOUR

SECRET SAM

"All of you kids. The lot of you. Come here. Front and center."

My father was doing what he occasionally did on a Saturday afternoon at home. He was summoning the runny-nosed children of the neighborhood, any within range, and telling them to stand in a line on our front lawn. He was producing a stack of Kleenex, neatly folded over, from the back pocket of his blue jeans and he was moving down the ranks, swiping away mucus with an impatience that made little heads bob. He was clamping the Kleenex around the nose of each child and ordering him or her to "blow."

"And blow."

"And *blow*."

"C'mon, *really* blow."

The neighborhood kids, who could not know my father as I did, would stare back at him bewildered at his gruff attention. But that was my father, fully in character. With every "blow!" he was establishing a new and tidy order, carrying out a hands-on job he'd invented for himself, inventing purpose where there had been nothing but the lazy aimlessness of a Saturday in the suburbs.

The nose blowing was a minor flash of impulse for my father. Most of his self-assigned tasks had about them (it seemed to me as a child) a grandeur of scope and motion and noise. Every weekend it seemed he brought home some new power tool—drill, skilsaw, saber saw, belt sander—all of them deadly serious in their gray metal boxes and, when brought to life, all the more serious for their loudness and potency. These tools made "the impossible doable," my father would say to me. Also, they would "slice through skin and bone like butter" if you didn't know how to use them properly. I had seen my father use his power tools to build a set of sawhorses and then, using those sawhorses, build all the other things he found necessary to build: his work bench where all his tools were neatly hung and stowed; a loft high in the garage rafters where the Christmas decorations were stored; a lidded toy box in the family room; a drying rack for the walnuts that grew in our backyard; the redwood deck shaded by its bamboo covering; a brick walkway around the deck with a crosshatched pattern taken from a do-it-yourself booklet.

There was nothing my father could not do himself, apparently, though rarely did he make much of an announcement about his day's plans. You knew they were underway because you woke up to hear, say, the creak of shingles overhead as my father erected a new TV aerial or cleaned the drain gutters. Or you would hear the big Dodge 440 station wagon *vrooming* to a stop in the driveway, my father back from an errand you didn't even know he was on, a load of planks sticking out far beyond the dropped tailgate, red flags flapping from their ends.

I surely wanted to be near the red flags and raw lumber, the loud drills and saws, whatever thing, whatever day, my father would make of them. But I also knew instinctively that all my father's relentless motion, like his blowing of the neighborhood's runny noses, was born of a mysterious impatience, an impatience always there within him. I could sense it as I stood by the bathroom door watching him flick clean his shaving brush, a sudden eruption of wrist snaps ricocheting water and soap against the mirror. I saw it from my seat at the dinner table as I observed him shake up a bottle of Wishbone Italian, another overkill of wrist snaps turning oil and vinegar to froth in a second. I could see it in the briskness and tight focus he brought to any chore, as if he were racing to meet a crucial deadline.

The trick for me was to probe this impatience continually, to make some estimate of it and to guess whether, at any given moment, it might become my enemy. I always needed to know whether the impatience was now very close to the surface, verging on anger, seeking a target, seeking a target in me. I did best to watch and listen from a discrete distance. I listened for certain words that signaled the all clear. I listened, for example, for the word "copacetic," slang from black jazz culture meaning "all right," a word that had found its way into my father's vocabulary by way of the Navy. "Everything here," my father might announce while hooking my bicycle chain back onto its sprocket, "now appears copacetic." And so it was.

I listened for sounds my father would make. The click of his tongue, for example, against the roof of his mouth, a crisp, hollow *shtok* of satisfaction that things were going according to plan. *Shtok, shtok, shtok,* he would say as he dropped ball bearings into the grooved race he had cleaned and greased. *Annnnnd shtok* as he dropped maraschino cherries onto a row of ice-cream sundaes.

I listened for the most welcome signals of all, the bad puns and jokey routines he invented for just the two of us to tell each other again and again. When he wanted to share a laugh with me, my father might borrow the cartoon voice of Yogi Bear and pronounce the two of us *Saaaamarter than the average Beers*. If he was in a particularly good mood, he might go into one of his Maxwell Smart riffs. *Sorry about that, Chief!* he'd say as he tickled my legs with a spray of the garden hose. My father found *Get Smart* hilarious: the Cone of Silence, the phone in the shoe, the spy in the file drawer, the way Don Adams passed through one secret door after another until the last one, clanging shut, caught him by the nose. Watching Max suck up to his superiors while bumbling his way to another victory over evil KAOS, my father laughed often and hard.

The impatience in him had made my father leave the Navy to test jet engines for General Electric in Ohio. "I hated life aboard ship. No sense being a sailor if you can't stand sailing." Impatience again had caused him to leave GE for California and Lockheed. "I had stars in my eyes for the technology. Dreams of the cutting edge." The jet engine, after all, was a creation of World War II, its future at best one of endless refinement. But the technology of the space race was fresh and without clear limits, the sort of blue sky work done in LMSC Building 104, massive and windowless, where my father reported for the first time in August of 1960.

He was put to work on a spy satellite project called Samos. Already well into development, Samos was to be many elements brought together. Samos was a powerful Atlas booster that lofted the satellite into space and then fell away. Samos was the orbiting space vehicle called Agena, a twenty-five-foot-long, five-foot-wide cylinder stuffed with guidance, tracking, and propulsion apparatus. Lockheed built Agenas assembly-line style for a variety of satellite projects. My father remembers his first glimpse at one as it lay in a work bay stripped of its outer skin, technicians buzzing around it. As his eyes played over the "accelerometers, gyro-guidance package, electronics for radio links, spherical tanks, hoses and tubes, rocket motor," he found Agena "a gorgeous little thing. Everything shiny, bright and clean. It just looked good."

Samos, too, was the spying payload mounted onto Agena. This payload was a kind of flying Polaroid camera plus fax machine. According to plan, once in orbit Samos was to shoot pictures of Earth's surface, develop its own film, convert the negatives into radio signals, then beam those signals down to the

Valley of Heart's Delight, where rooms full of workers at consoles would unscramble them back into pictures.

Samos, in short, was to be a complex system made up of many different systems, and this made Samos a perfect example of what aerospace engineering was getting to be wholly about in the early 1960s. My father had understood this when he leaped at Lockheed's invitation to come join. He was eager to be among people who were pushing forward "systems theory," people who were practicing and perfecting the promise of "systems analysis." Now that the government was in the business of commanding into being grandly ambitious technological goals—nuclear arsenals, moon rockets, spy satellites—the need had arisen for a new approach to invention, a science that concerned itself with the breaking down of any such blue sky goal into myriad smaller ones. And, then: figuring out how to mesh all those discrete goals, once achieved, back into a functioning whole. Systems engineering was that science, and my father saw his future within it.

My father, as it turned out, was not to spend any more time in the corporeal presence of the gorgeous Agena. Nor would he be allowed any access to the camera payload that made Samos truly cutting edge, nor would he be invited into the rooms where the spy pictures were eventually received and processed. Within the system of Samos, it fell to others to lay hands and eyes on the actual machinery and to handle the true final product, the images collected.

My father's function on this, his first Lockheed project, was to be an Orbital Test Planner. This meant that my father's precise duty within the whole was to sit at his desk and imagine the satellite, whirling in space, as a mathematical abstraction. Lockheed needed to program the flight of Samos in such a way as to best conserve power and film aboard. This involved finding the optimal times to turn certain switches on or off, and this could be expressed as an equation on paper. My father was paid to make numbers slide off his slide rule, numbers which would eventually be placed in the hands of other men whose duty was to sit at the controls that guided the actual Samos satellite through space. By that point in the process, some little change in protocol could easily have made my father's numbers irrelevant, and so he never knew whether they were used or not. By that point in the process, however, Lockheed had already moved him to another desk where he was asked to imagine a different satellite, whirling in space, as a mathematical abstraction.

After some years of this Lockheed would grant my father a new phase in his early career, naming him a Systems Test Engineer. The work proved no less abstract and still the gorgeous Agena remained shrouded from view. "Most of my day," my father remembers, "was spent looking at columns of numbers, data sheets, comparing specs with hardware and deciding whether numbers were good or bad." His work was not done until the bad numbers had been defeated and replaced by good ones, until he could explain why the good numbers had come to appear to him shiny, bright, and clean, gorgeous little things in their own way. He would write all these good test results into thick binders full of graphs and tables. When enough such documentation had been accumulated, when the equation had been sufficiently solved, the Systems Test Engineer was to play salesman. He was to present his test results, well before the actual launch of the satellite, to help try to convince the government that Lockheed had fulfilled its contract to date and that payment therefore was due. This "DD250 Sell-Off," as such meetings are called in the aerospace business, went well enough my father's first time out. The officials of the Air Force who were buying on behalf of the American people were impressed by the sparkle in my father's numbers. They authorized the release of some millions of the hundreds of millions of dollars budgeted in total. And then, under the harsh light of fluorescence, my father returned to poring over other data sheets in search of new numbers good and bad.

If such a life seems an unlikely choice for an impatient young man, my father embraced it without reservations. A few weeks on the job convinced him that a successful blue sky career, which he surely wanted, meant tackling abstractions with the same can-do spirit he would bring to flying a jet or fixing a car. He understood, too, that for reasons of national security (and what could be a more excitingly important reason than that?), his work must be, in Lockheed language, "highly compartmentalized." Whenever the upper reaches of the corporation handed down to him a new, cleanly circumscribed task, he would be told only what he needed to know to do his job and nothing more. By the same token, he was to tell fellow workers only what *they* needed to know. Compartmentalization ensured that each worker was allowed to feel his part of the elephant but never to see the entire beast in the light of day. This made it possible to harness the skills of thousands in order to achieve a single blue-sky goal while reserving for the very few—those who conceived and controlled the overall project—answers to the big questions: How will this ultimately *work?* What is this ultimately *for?*

My father accepted this culture of compartmentalization on terms an aspiring systems engineer could understand. Secret information was like any other volatile substance, he reasoned. A system built to manage its flow must be designed for maximum control and safety. What my father liked about this system was that he could imagine a path of progress for himself within it. He understood that the honeycomb of compartments was organized by a hierarchy of security clearances, at bottom Confidential, then Secret, and, highest, Top Secret. Clearances were known as "tickets" at Lockheed. To get his first ticket, a Secret-level clearance, my father had told the government investigators all they wanted to know. He had listed every place he had lived over the past ten years and he had sworn to them that his family was clean of foreign nationals. My father was only too happy to oblige. What had he to hide? Nothing, and everything to gain.

The vetting finished, my father had been given papers to read and sign, understandings that if he spilled certain details to the wrong person, if he breached the walls of his information compartment, he would land in Leavenworth prison. He signed the papers and thought to himself, "I've been admitted into the inner sanctum." Of course the innermost sanctums required a

Top Secret clearance, but my father had no doubt that if he proved himself, he would eventually earn one. My father was impatient to prove his competence within Building 104, where competence proven won this reward: further progress into a realm of hidden work, a career track leading to the heart of, as Lockheed language termed it, the Black World.

I learned early to study my father's face as he came through the door after his Lockheed workday. If his eyebrows were where they should be—at rest on a line-free forehead—there was every chance of the usual Dad, the wisecracking Dad, who would want to know all about how "life around here" had gone that day. If he gathered up my mother in a languorous kiss and called her one of his nicknames, if he said to her *What's new, Scrappy?* as he pulled from the cupboard the glass bubble for mixing martinis, if he filled the mixer with ice and liquor and stirred this all around with the glass wand, if he lifted a child into one arm while he loosened his tie with the other hand and then took a lip-smacking sip of his drink, if he did such things, then prospects for the night ahead were excellent. He might even be coaxed after dinner to transform himself into the Hairy Umgawa, the monster who wrestled all comers on the shag carpet of the living room until, inevitably, he lay panting and defeated under a pile of children.

But what if, when my father came through the door, the eyebrows were not where they should be? What if a critical mass of lines had gathered on his forehead and pressed the eyebrows together and down? What if he stepped through the door to the commonplace sound of a pot clattering or a baby crying and those eyebrows darted low even as the eyes seemed to widen and show too much white? These were indicators that my father was this night, at some point, likely to erupt in rage. I remember many dinners that went from happy chatter to grim conflagration in the instant it took a child to knock over a small glass of milk.

Cripes sake! my father would yelp, shoving himself off his chair, grabbing for a dishrag at the kitchen sink, flinging the rag at the table, at all of us, it seemed, just for being at the same table where a glass of milk accidentally had been spilled. He'd sop up the mess in a matter of seconds, but the acrid mood he'd created would settle over everything, saddening my mother, who had expected a better reward for her dinner making.

Whenever my father's impatient anger would find a target in me, my day would disintegrate, laid flat by a blast of browbeating. The touch-off might be the skid mark I made on the driveway with my bike tire, or the screwdriver I had forgotten to return to its hook on the wall, or the grass clumps I had unwittingly tracked into the house from the backyard. It might be the unmade bed that caught his eye at eleven A.M. on a Saturday morning or the noisy tussle I was having with my little brother on the same shag carpet where the Hairy Umgawa had been last night. Whatever did set him off was likely some bit of disorder that hadn't much bothered him the day before, but this day his impatience had the better of him and I hadn't taken proper precautions and therefore he now was standing over me, his eyes with far too much white in them, his face inches from my own. *Useless ninny!* he was shouting. *Giiiyaaad!*—a yell that trailed off into a gagging sound. That I did not think clearly, that I did not make myself useful, that I was a whiner in the face of duty, these were the contemptuous accusations my father the engineer would level at me when he was really worked up. *Giiiyaaaad you're useless! Useless! Have you a brain in your skull?!* I would find it impossible not to cower, not to cry, so I would cry and that only ever redoubled the onslaught—*Stop your pathetic blubbering!*—until my head ached and the world was red-tinged by tears and everyone within range was utterly miserable.

And so to prevent any such scene from suddenly happening, I watched, I listened, I fine-tuned my powers of surveillance whenever possible. If I had a clear view of my father's face, I studied the forehead, the eyes, the corners of the mouth that might tighten and dip. If I was coming upon my father from behind, I observed the neck. Was it rigid? Did the neck seem to retract at the sound of my feet and my voice, cock just a bit to one side, making it easy to believe that the eyes I could not see were now clenched shut against my very presence? If his legs were all I could see, sticking out (as they often were) from beneath an automobile in the garage, the thing to do was to listen for his grunts, the tone of them as my father grappled with the repair job. Some grunts, the favorable ones, were rounded, open *uhhhs* resonating with satisfaction at progress made. As often, though, they were bitten off growls of frustration, the surest sign that I should not say whatever I had come to say, that I should move on and not let my father know I had ever been there.

Sometimes, that option was denied me. "Dave! I need your help here!" my father would call out. Perhaps he had dived, fearless, into a jumble of wires inside the living room wall, or he had winched the car's impossibly complicated guts straight out of its engine compartment, or he had opened the back of a television set and was probing that inscrutable landscape with yet another strange and new tool. At times like this my father who could do anything would tell me to stand by him and hold a flashlight beam on the exact spot where he was performing his mysterious manipulations. At these moments, when I was granted so close a view of my father's secret powers, their source remained tantalizingly beyond my understanding all the same. My father was not one to teach as he worked; impatience prevented that. "Damnit, Dave, put the light right *here,*" he would say as my attention inevitably atrophied and my aim relaxed. I could have tried to learn what he was doing by asking questions, of course, though I knew that if I asked one too many at the wrong time, there was always the chance that my father's impatience might jump like electricity from the task that frustrated him and over to me. And so, at times like those, I watched the lines on his forehead and listened to the tone of his grunts while asking little, my father offering little, father and son collaborating on the day's important project, each of us operating strictly on a need-to-know basis.

Samos not only worked as intended, the satellite became America's most famous spy for a time. As any layperson could read in the *Los Angeles Times* in October of 1960, Samos was meant to keep "this Nation informed of vital military installations and build-ups behind the Iron Curtain." There was no attempt to hide the project's Cold War nature. All the world knew that the Soviets had in May of 1960 shot down the U.S. pilot Gary Powers in his high-flying U-2 reconnaissance airplane. Now it was publicly understood that Samos satellites would carry on the espionage from a height truly out of reach, flying through what Eisenhower had declared the "Open Skies" of outer space.

By summer of 1961 a Samos satellite had made some five hundred passes over the Soviet Union. The pictures Samos 2 took helped give a lie to supposedly "superior" numbers of Soviet nuclear missiles, a superior strength that the Kremlin and the Pentagon alike had been proclaiming, a "missile gap" that Kennedy had sounded alarms about in his presidential campaign. When Kennedy and his brass got their pictures back from space, such scare talk immediately cooled, and Americans breathed easier. Samos, it seemed, had pulled off the gambit that good spy stories turn on, the unmasking of the enemy's bluff, the stealing of secrets to shift the balance of power back to the good guys.

My father felt glad, his first year at Lockheed, to know he was part of this high drama with a national pride ending. This seemed a more romantic start to a career than had he been assigned to, say, Discoverer, a Lockheed satellite project underway literally next door to Samos. Discoverer, which had been fraught with fizzles for years, was said to carry into orbit capsules containing mechanical "mice" wired for biomedical data or sensors for measuring space radiation, the sorts of Science for Peace experiments that Ike and JFK and Khrushchev all declared they wanted of their space programs. After a time in orbit, Discoverer would drop its capsule back toward Earth, an airplane trailing a scoop would catch it out of the air, and humanity would have gained some new bit of knowledge about ourselves or the cosmos. That was how Discoverer workers would explain it to my father if he happened to chat with them on the way to the doughnut wagon. They told him just what America had been told, that Discoverer, for all its troubles, was the friendly, civilian face of America's space effort.

What my father found out some years later—and America many more years afterward—is that he had been lied to on the way to the doughnut wagon. Samos had not been America's premier spy satellite after all. Just a few months after Samos's first success, friendly Discoverer, its bugs ironed out, began dropping spools of exposed film, spy photos with sweep and resolution far better than Samos could ever achieve. While Samos certainly made a contribution, it was Discoverer that decisively debunked the missile gap, and while the Samos program was curtailed in 1962, Discoverer spy missions, under the code name CORONA, continued on in Black World secrecy until 1972. Looking back to his very first months at Lockheed, my father remembers strangers coming and going in the halls of the Discoverer project, officious men he might have guessed were intelligence officials from Washington. There were enough clues at the time, my father now sees, that he probably could have guessed that Discoverer was for spying, and that his coworkers on the project next door were lying to him as they were required to do. The reason he did not guess this is that he would not allow his mind to wander beyond the compartment Lockheed had assigned it. My father knew that he did not need to know.

All this about Samos and CORONA and my father's relationship to the projects I say because my father has spoken freely about it, a luxury allowed him because Samos was so well publicized, and CORONA, just recently, has been declassified. There is very little else I know about my father's projects during his thirty years at Lockheed. I know that he continued to punch a time card at the Satellite Test Center for at least seven years after Lockheed hired him, and that after Samos he worked for a while on Vela, an unclassified satellite capable of detecting a Soviet nuclear detonation. I know that one day in the early years of his career my father announced to us that he had managed to wangle a transfer to a particularly exciting Lockheed project, a nuclear-powered rocket to the moon called RIFT. I know he and my mother attended a big going-away party for him on Friday, and that the headlines in Saturday's newspaper announced that the nuclear rocket program had been canceled, which meant that my father resumed his job, on Monday, at the STC.

I know that my father eventually did receive a Top Secret ticket allowing him to pursue, in some form, the more interesting work he coveted. I know that some years after that he was taken aside and told that Top Secret was not, in fact, the highest security clearance in the company after all, that Lockheed had long awarded Special Access clearances for higher level work on projects so secret, they were funded from an unlisted "black budget." The very existence of this system of secrecy had been kept secret from my father until he was deemed worthy to join it. I know that my father, having been invited into this innermost sanctum, was given a yellow checkerboard badge imprinted with several numbers, each of them signifying a different Special Access granted. I know that over the years the numbers on the yellow badge accumulated, and as they did there was less and less that my father could tell his family about what he did for a living.

The system of compartmentalized secrecy at Lockheed was self-regulating in more than one sense. Workers were expected to check each other's badges and to turn in any fellow worker who spoke too freely or handled secret documents improperly. As well, any ticketed worker at any time might be administered a polygraph test by the company's security people. "If you found yourself flunking polygraph tests, your clearances were rescinded," my father explains. "Your ticket was pulled, which made you an unemployable engineer at Lockheed. You'd go to NASA where the work was completely unclassified and deadly dull. You wouldn't get money and promotions; the money and promotions were in the highly classified jobs. Security clearances were precious commodities. They were merit badges that opened all kinds of doors for you."

Whenever my father was polygraphed by his employer, he naturally would submit willingly, having nothing to hide and everything to gain. Had he revealed details of his work to anyone outside the proper compartments? No, my father could always answer truthfully.

At the dinner table, his young son would also ask him questions. "What do you do, Dad?"

"Welllll . . ." A question like that always invited a long pause. "At times in the past my work has involved me with satellites."

"I know, but are you working on a satellite now?"

"Hmmmmm." A pause. "I am not able to give you a yes or no on that one." Another pause. "Let's just say that I'm helping to troubleshoot a very complicated piece of equipment for the government."

"Is it something that goes into space, Dad?"

My father would chuckle and he'd cock an eye at the ceiling for a bit. He would be weighing just what he could say and what he could not, what he might tell his nine-year-old son that might not show up later as a damning twitch of the polygraph's needle. At my family's dinner table, Lockheed was always listening in.

"I'm not really at liberty to go into the specifics," was the answer my father often gave when my questions about his work invited the least bit of specificity. I soon learned to stop asking.

To have a father who worked on secret projects, to have a father who was himself a secret project in our midst, did not seem to us a deprivation. It meant to us that we were a promisingly modern family. The old notion of father as civic actor, the man who displays his competence in the public realm, the doctor or deacon or grocer on the corner, had, as everyone knew, given way to Dad as Organization Man. Now that fathers no longer lent their competence directly to the community, now that they honed and displayed their competence within the closed system of the corporation, how natural it was that our father, a most modern Dad, worked within the most closed of systems. We did not think it strange that we never saw my father's desk. Whenever we asked about it, my father would shrug and tell us he worked in various cubicles he purposely kept bare of pictures or knickknacks. "I don't believe in making my place of work homey," he would say, and so we never gave him presents for the office.

Where we lived, there were many other fathers like ours, men who disappeared into windowless buildings every morning and who, when they returned home, spoke vaguely about what they had done that day. When they gathered together on the weekends, they stuck close to whichever topic had drawn them together, the car being fixed or the fence being built or the cement pathway being poured. It was obvious even to a young boy like me that these men shared something, a sensibility owing to technical training and very much prone to creating purpose in the midst of an otherwise lazy, aimless Saturday in the suburbs. And I believed that whatever it was they had in common in their jobs, whatever it was they preferred not to discuss much with each other, it must be important in a way that the work of a grocer or deacon or even a doctor could not be. "If and when the Soviets ever launch their missiles," I can remember my father telling me more than once when I was quite young, "you had better believe that Lockheed's right at the top of their list. We'll be the first to go. Lockheed, this house, everything in a fifty-mile radius. *Kablooey.*" Because my father always looked more winking than sad as he said this, I inferred that beyond the horror of his statement there lay reason to feel good, even proud about it. If we lived at ground zero, it must be because we were special. Perhaps I would never know what my father did at Lockheed, but I could rest assured it was important. That meant (should the enemy missiles not come) a bright future for me and every member of my family.

Did I imagine myself, then, in that future, taking up the

important work of my father? I have no memory of ever expecting that. After all, how does a young boy imagine himself following in his father's footsteps if those footsteps lead into blackness? How can a secret systems engineer pass onto his son the tenets of his profession? With no picture in mind of what my father did at Lockheed, I turned my gaze toward what he did at home. But even at home, my father managed to make his projects secret ones, hidden behind his wall of impatience. I wonder now how my formation would have been different had I been a less cautious child, had I bulled away at my father until he swallowed his impatience and slowed his work and included me in the thinking behind it. I know there were times when he was a great explainer of the Whys and Hows of the world. Next to a bonfire at the beach, he would explain why flames crackled. Between innings of a baseball game on television, he would explain how a curveball curved. When we were together in the cockpit of an airplane, he would spend all the time it took to help me to understand what made the wings lift us into the air. But those were times when he was relaxed and inviting of questions. Those were not the times around the house when he was all action, seeming to move *among* his family rather than in any way *with* us. Those were not the times when my father wielded his can-do-anything competence with an urgency that drew me toward him, but with a fierceness, too, that kept me at bay.

At some point when I was ten or eleven or maybe twelve, my father and I settled into an unspoken agreement that I was not "handy." I had not demonstrated an appetite for the technical, the making of purpose in an empty day. I had not done much with my Erector set, had never finished my plastic model of the U.S.S. *Constitution,* did not seem to want to do more than hold the flashlight for my father whenever he was in action. I was not a handy boy, and so I filled up the lazy aimlessness of a Saturday with dreamy play. Most often, I constructed my fantasies around the theme that obsessed American popular culture at the time: secrecy and spying. James Bond was too racy for us kids to see, but his appeal had been well commodified and kiddified by the mass marketers. Between the Saturday morning cartoons there were ads for spy toys. "Six Finger! Six Finger! Man alive! How did I ever get along with five!" went the jingle for a toy gun that looked like an extra finger in your hand, the very sort of weapon a secret agent on TV might use. I got one of my own and, as well, one of those gizmo-stuffed attaché cases that every spy carries. The Secret Sam case, as it was called, contained among other things a breakdown rifle with a scope and a hidden camera that allowed me to take pictures of playmates without them knowing it.

I took as my models the dashing men of cool competence and action I saw on *Secret Agent* and *The Man from U.N.C.L.E.* and *Mission Impossible.* These television shows seemed to grab my imagination far more than they did my father's. But then, my father was intimate with the evolving nature of Cold War espionage. He knew that by then the ones who guarded and stole the nations' secrets were those who spent their days studying data sheets looking for numbers good and bad. He knew that from here on out, the most successful secret agents would be robots in space, built by compartmentalized workers who lived in the suburbs. Perhaps that is why my father preferred to watch *Get Smart,* to laugh as silly Max passed through one secret door after another until the last one, clanging shut, caught him by the nose.

Saturday would come and my father would be off in the Dodge 440 on an errand to pick up what he needed to stop the washing machine from leaking or to run a gas jet into the family room fireplace or to install a new sprinkler in that area of the backyard lawn that was turning brown. He would be away from the house, and so I would climb from one of his sawhorses onto a niche in the chimney and from there I'd pull myself onto the roof. I'd be playing the theme song to *Mission Impossible* in my head—*dun ta dun DA da, dun ta dun DA da, na na NAAA, na na NAAA*—as I stole across the shake shingles, making my way to the other side of the roof, below which there lay a neat mound of grass clippings from the morning's mowing.

From this high vantage I would look across the site of all my father's projects, taking in our house and yard and garage and driveway, imagining this to be an enemy compound I had been sent to infiltrate. Below me, where the clipped grass was piled, I would imagine two Soviet sentries standing guard over the compound. A man had to be certain and quick, a man had to know exactly what he was doing to take out two Soviet sentries at once. I would leap off the roof into the grass and slice the throats of my enemies, both of them with one smooth, expert stroke. And then I would lie in the fresh-smelling grass for a moment, letting the spy music play through my head. I would lie there drifting on the edge of my father's Black World.

FIVE

LOST IN SPACE

Nicky Giannini, who lived in the yellow three-bedroom at the end of the cul-de-sac, was my best friend. Whether it was time to ride our Stingrays, dig roadways in the dirt for our Matchbox cars, sit in the branches of a walnut tree, or catch grasshoppers in a jar, the two of us instinctively knew it, did it. We roamed the sun-drenched universe of our neighborhood as if guided by the same map of impulse.

Only on those rare days when Mrs. Giannini had us inside to play, on those days too hot or too rainy to be out in the universe, would I be reminded that Nicky came from a world strange to my own. Mrs. Giannini was the only mother on the cul-de-sac with gray hair, which she swept back in an unfashionable pile. She wore old lady's eyeglasses, silver-rimmed and cat-eyed, and she began her cooking of dinner *early*, on some days even before lunch, filling the house with tomato sauce smells. The house she kept was far more precise and muted than my mother's, devoid of the flash, the daisy yellows and pumpkin oranges, of my mother's taste. In the Giannini living room, where no children were ever allowed, there was a polished upright piano, and a fancy couch with clear plastic over it. In the family room there was an elaborate (and said to be dangerous) machine just for pressing clothes; the rest of the space was filled with vinyl furniture, a TV on casters, and a braided oval rug onto which Nicky and I would dump his huge box of Tinkertoys.

There was, as well, a hi-fi that Mrs. Giannini liked to play much of the day. Her records were ones I could not imagine my mother and father owning. A man named Dean Martin who sang with a *who cares?* leer in his voice about the moon being a pizza pie. A man named Jimmy Durante, big-nosed and clownish on his album cover, who rasped out baby words like "Inka-Dinka-Do." My parents listened to singers who made a serious attempt to do well. They played "sound tracks" like *My Fair Lady* and *Camelot*, and my mother particularly liked the earnest harmonies of Peter, Paul, and Mary. Mrs. Giannini did not have anything like Peter, Paul, and Mary in her collection.

Nicky's father was not only older than any other father I knew, he was much more formal, a man who wore pressed white shirts even on Saturdays and who expected a true Italian supper—pasta *and* meat—when he arrived home after his day as a supervisor at the phone company. Nicky was the youngest of three, which made life around his house seem all the more exotic to me. When Nicky and I were nine or so, his sister, Celeste, was primly nearing the end of high school. About Nicky's brother, Joe, a few years younger than Celeste, there was nothing prim at all.

Joe loved to make his eyes twitchy and then lock them onto yours and let out a high, deranged giggle. "You see these weeds?" he'd say to me as he chopped away with a hoe at some dandelions in a flower bed (Mr. Giannini required Joe to keep the lawns and garden in perfect condition). "You see 'em? They're the Red Chinese" (twitchy eyes, mad giggle) "and I'm an H-bomb!" His red-blond hair already thinning in high school, his small frame sinewy from lifting weights to rock 'n' roll in his garage, Joe looked nothing like the rest of the Gianninis because he had been adopted from Ireland, a fact he celebrated by hanging Emerald Isle tourism posters all over the room he shared with Nicky. Joe did not have a best friend to help him navigate afternoons on the cul-de-sac, and so he invented his own amusement. He searched out black widow spiders in dark crannies, keeping them in a glass jug, feeding them flies and urging Nicky and me, whenever we were around, to come and see his nesting pets. "After they mate, know what the mother does to the father?" (twitchy eyes, mad giggle). "Sucks his juices out!" Joe's hobby fit well with his love of the horror films that played on television at three-thirty in the afternoon. My mother did not want me watching such disturbing fare, but Mrs. Giannini was less vigilant, happy, it seemed, just to know her boys were close by but out of her way as she labored over the night's rigatoni and roast beef.

"Nicky!" Mrs. Giannini would shout down the cul-de-sac in a screechy voice that had no trouble catching up to us on our Stingrays or in the branches of a walnut tree. "Nicky! You come inside for a while. Bring Dave. It's too hot! Come inside and watch TV!" We would be summoned inside of Nicky's world and Mrs. Giannini would turn off Dean Martin and Joe would find the monster movie channel and over the next few hours Nicky and I would watch, quivering and rapt, while Joe giggled and Mrs. Giannini cooked. We would watch Godzilla lay waste to a city or the Blob melt unsuspecting lovers or the Wasp Woman seduce and kill or the Tree Monster crush those unfortunate enough to dally in its branches. When the movie was ended and dinner was ready and Mr. Giannini was due in the door at any moment, Mrs. Giannini would send me home, my knees weak and my head aswim with alien forms of life.

Behind every same door, within every same floor plan, a world unto itself. This, to my tribe, was the appeal of the detached single-family residence in the blue sky suburb. The sameness of those houses reassured us that, despite the closed curtains, all of us were similar people. The detachedness of those houses meant the possibility of removal from unwanted variables. A blue sky child was to be given sun and space, his mother nearby, the companionship of siblings and other children like him, was to be given religion and manners, was to be given a safe and managed universe within which to define himself. That was our faith. That was thought to be enough, in the middle 1960s in neighborhoods like mine, to ensure our tribe's regeneration. The children, so incubated, would acquire the necessities of white collar success, the same drive and propriety that had won each house, each "nice home," for their parents.

Critics of the suburban form have long argued that such faith made us our own dupes, that by fleeing the difficulties posed by the cityscape, the public realm, we opted for the soft life of private consumption. They say that a family together alone in its tract home made the perfect patsy for the spine-weakening seductions of TV. When I hear this I find myself thinking back to how my father and mother made me feel about the television set that sat at the center of our house. They conveyed toward "the tube" (as my father always called it) a deep suspicion, even resentment, of its presence, its power, in our midst. "Insipid!" "Utter waste of time!" "That damn boob tube!" "Tripe!" my father would fume as he strode through the family room on his way in or out of the garage, his cloud of contempt temporarily blocking our view of *Bewitched* or *I Dream of Jeannie* or *The Addams Family*. Mom's quiet, shooing suggestions that we find something else to do were just as damning. What Mrs. Giannini would call to Nicky—*Come inside and watch TV!*—my mother would never say to me.

I knew that the pastor of Queen of Apostles told us not to covet, and then on television folks dressed up as fools and squealed with greed at the feet of Monty Hall. I knew the nuns told us to tell the truth, and then the TV ads brazenly lied about soup and soap and cigarillos ten times an hour. In the morning, death was something somberly sacred, wrapped in incense and intonations at the funeral Mass I'd serve as an altar boy, and then, that night, death happened between corporate jingles, a plot device on *The Rat Patrol.* My father and mother did not fail to remind us of such contradictions, nor, by the looks on their faces, of where they stood. My father's continuous critique of the tube followed several lines: That most television was a threat to our intellects ("mindless"). That most television was work poorly done ("junk"). That most television was the domain of people on the make ("hucksters") who did not care for our best interests. My father made it known that there was a monster in the universe beyond our cul-de-sac, a great crassness that wanted into our home, our world unto itself, via the tube.

As a family, we nervously co-existed with the television, for it was understood that television, like temptation itself, could never be banished from our existence. We did allow as how there was such a thing as a "good" show. My mother considered Captain Kangaroo "good" for small children. And, as well, some nights my parents would actually wake me from my sleep to watch something "good" on TV, a documentary, for example, about life onboard an aircraft carrier. When my father laughed at *Get Smart,* I understood that, while the show was not necessarily good, its small value lay in the fun it poked at much worse TV. *Hogan's Heroes* could never be good because "there's nothing funny about a concentration camp," my father said, and so I watched that show with the sense of self-loathing my father intended. The monster movies I saw at the Gianninis' were certainly nothing good, crass attempts to shock with no point to them.

If their children had to have monsters, my parents would allow us the sort found on *Lost in Space.* Here was family fare about a blue sky family of the future, the Robinsons. John, the dad, was a scientist. Maureen, the mom, well, her impeccable credentials were that she was June Lockhart, Lassie's mother in a silver jumpsuit. Dad and Mom and daughters Judy and Penny and son Will had set off in their flying saucer with their handsome pilot, Don, to begin a new life on a planet in the Alpha Centauri star system, only to be sabotaged by the villainous stowaway Dr. Zachary Smith. Now the group found themselves crashing into one uncharted planet after another. The monsters they ran into were so campy (one was a walking, talking carrot) that I cringed at the cheesiness even at my young age (eight when the series started in 1965).

And yet I was drawn to *Lost in Space* like no other show, pointed all week toward 7:30 on Wednesday night. The cliff-hanger endings no doubt made me want to tune in again, as did the fact that most stories revolved around the perils of Will, who was smart and serious and just about my age. But thinking back, I see that the very premise of the show, a dark one, must have exerted a visceral pull on me. *Lost in Space* was gloomier by far than its contemporary, *Star Trek,* which predicted a future in which Earth's inhabitants, having reached peaceful higher consciousness, would join with other planets under galactic government. Kirk and his crew, an integrated society with even a Vulcan aboard, were men (and a few women) off at work: a smooth-running corporation of technical specialists flying through space with the ongoing support of the Federation.

But the poor Robinsons. Earth had become overcrowded and, as a result, they were white middle-class folks who'd volunteered for a transfer. Now they found themselves every week alone together in the ultimate single-family detached residence, forced to invent over and over their own haven in a heartless cosmos. Mom would be tending the hydroponic garden, and Dad would be combing the new neighborhood in the van (with tractor treads and spinning radar dish). Life would seem, finally, to be improving for the Robinsons. But then some monster would invade their storyline, usually with the collaboration of the devious and cowardly Dr. Smith. Now it was time to see how a family performed under pressure. Tomorrow's family together alone, beating back monsters wherever fate had landed them.

My mother, as she often said, got "a kick out of" Mrs. Giannini, with whom she shared a warm friendship. But my mother, who had four young children, found more in common with Mrs. Williams, who had five and who lived two doors down. The two mothers would sit in the dining nook of our house some mornings, an ashtray between them filling up with cigarette butts while babies swarmed around their feet. They laughed a lot, I remember, shaking their heads as they blew smoke up at the ceiling.

At that time it was possible, of course, for a mother like mine to be home with her children all day, every day. It was possible, too, for a son like me to find his mother an enigma precisely because of what her being at home all day entailed. The work of my mother was an invisible timetable of needs to be met, carpets that had to be swept before company came, windows that had to get washed some time over the long weekend, bathrooms that had *already* gone too long without cleaning. Every mealtime balanced on the point of her focus, every crying child was her problem to address. What did a modern mother do? Nothing a son could imagine doing, could imagine *wanting* to do. He only knew that a mother was required to be simply always *there*, a *there-ness* that would make our tract home world carry on in its existence the way the *there-ness* of the sun made day after empty day exist for the benefit of playing boys.

There were many days when my mother would begin to feel taken for granted, would want to pierce her children's indifference to her *there-ness*. She would mope and chip at us, telling us that, for example, "If I were laying there dead on the kitchen floor, you'd all just step over my body on the way to the refrigerator for a Popsicle." Some days, her lament—"Just once I wish someone would help out without being asked . . ."—would rise to excoriation. *Damnit to hell!* my Catholic mother would scream. She would scream, but never as loud as Mrs. Williams, whose diatribes traveled through open windows and all down the cul-de-sac.

The possibility of being home in a blue sky tract home with

her children all day, every day, was something my mother had eagerly pursued as a modern opportunity, a freedom, the freedom to mother well. Yet one of the things she liked to laugh about with Mrs. Williams was the *im*possibility of *not* being with her children all day, every day. That was a perverse function of the "functional" design of the homes and the neighborhood we inhabited. The "open interior" floor plans gave a child six ways to find Mom. The hollow core doors closing off the "master bedroom" and bathrooms were sieves for the sound of children's demands. For a mother to leave her home and stand in the front yard was to announce her presence to every child like a stuffed exhibit in a museum diorama.

Sometimes my mother and Mrs. Williams found refuge together by hiding in the plywood playhouse in the Williamses' backyard, but the cigarette smoke rising from the cut-out windows would give them away. "Seen my wife?" my father asked Mr. Williams one Saturday.

"She's in the playhouse with Mary Ann electing the next Pope," came the answer.

One time my mother climbed into the branches of the backyard walnut tree, but a child soon stood by the trunk looking up at her. My father bought her a bicycle with a basket on the front, but what destination, in the boundless galaxy of subdivisions that surrounded our own, was there for a mother to pedal toward alone on a bicycle? I remember her riding mostly to Queen of Apostles school at lunchtime, the basket full of neatly bagged hot dog lunches for her children.

"Filiarchy" is the scorning word William H. Whyte, Jr., author of *The Organization Man,* used to describe the universe I inhabited, the baby boom subdivision where a middle-class mother is too much in thrall to her children, too willing to be at home drinking coffee and smoking with another mother simply because the two mothers' children were at their feet playing together. "It begins with the children. There are so many of them and they are so dictatorial in effect that a term like *filiarchy* would not be entirely facetious. It is the children who set the basic design; their friendships are translated into the mother's friendships, and these, in turn, into the family's."

He was writing in 1956. In 1963, oppression by filiarchy was thoroughly denounced in Betty Friedan's *The Feminine Mystique*, a book which (as Barbara Ehrenreich has noted) drove straight at the middle-class fear of waste and sloth and stasis. The college-educated mother was said to be "infantilized" by housework and child care, by work doable by even the "feeble-minded," and so such work should be hired out, in the name of efficiency, to lower-class women. To be home all day, every day, with your children was to waste your college degree, to cede public life to men, to let your character go to mush until you had been reduced to nothing more than a trapped and passive consumer. Friedan's manifesto, which powerfully resonated with millions of suburban women all over America, did not penetrate the walls of my home, my world. Had my mother read Friedan, which she did not, she might have found some of her frustrations given a voice. But I doubt that she would have agreed with the prescription offered her for those days when she moped and screamed and felt taken for granted by husband or children or both. To have worked outside the home while her children were very young would have meant leaving her post, would have struck her as running *counter* to the ideal of self-reliance that our family wanted to see in a tract home life. My mother hiring a servant while she earned a paycheck might have equalled efficiency in the macro-ledgers of a Betty Friedan, but that outsider invited into our private world would have signified, to my mother, a snooty dodge of the scut work that every other blue sky mother in the cul-de-sac tackled every day. It even may have made her look greedy for things the extra money might buy, announcing herself to be truly an escapist consumer.

My mother preferred to place her faith in life under filiarchy. She preferred to vent when she felt too much taken for granted—*Damnit to hell!*—and then to laugh this off the next day with Mrs. Williams, the two of them blowing smoke at the dining nook ceiling, passing funny comments on the strangeness of the respective planets they'd come to inhabit, neither woman willing to say she was marooned.

Recently I have learned that two rules governed the crafting of any *Lost in Space* storyline. One was: *Mother must never be in jeopardy*. The other: *Mother and Father must not touch one another.*

These guiding principles were related to me by June Lockhart and by Paul Zistupnevich, who was for thirty years the closest assistant to Irwin Allen, creator of *Lost in Space*. Irwin Allen, who also produced *Voyage to the Bottom of the Sea, Land of the Giants, Time Tunnel,* and, for the big screen, *The Poseidon Adventure,* and who died at the age of seventy-five in 1991, was not a man with elaborate theories about how to speak to the psyches of suburban children of the space age. He was, as Zistupnevich admitted with some exasperation, a man who didn't much care if his stories made sense at all. But Irwin Allen—or to be more precise, Irwin Allen in concert with network executives—did expect the space family Robinson to adhere faithfully to the simple two rules. Indeed, by Paul Zistupnevich's recollection, there was but one exception to the rule that *Mother must never be in jeopardy*. That was an episode when "Mother went out to rescue Father when he was floating in space and she managed to get him back." Other than that, she was to be kept from the center of action out of concern for a young audience who needed to count on Mother always being safely *there*.

And to prevent that safety from being intruded upon by even the first signs of dangerous desire, there was rule number two. June Lockhart told me: "In talking about our characters, Guy Williams (who played the father) and I felt they should be a loving husband and wife. We were very surprised, later in the first year, when word came down from CBS that he was not permitted to touch me at all. Because in the pilot we were very demonstrative and affectionate and there were mother-and-father-kisses and you felt this was a very viable relationship. The word was that it embarrassed children to see their parents hugging and kissing. Even when it came to him giving me his hand to help me down off the last step as we got off the spaceship, the word was: 'Don't touch her!' "

Even more than the first rule, this second one Paul Zistupnevich found "kinda kooky," as did all the cast members. Lockhart said that she and Williams did the best they could with "longing looks," but it was written that the blue sky marriage of the future was to be all chaste efficiency in the eyes of the children. Zistupnevich, who designed the costumes for *Lost in Space*, dressed tomorrow's mother in material that did not invite hugs, firefighter's asbestos. Her clothes "were so stiff you could lean them against the wall and they would not collapse. You didn't have to worry about bust points or anything. If you shaped them, the shape stayed there."

Lockhart told me she hadn't expected such constraints when she took the role. She remembered being thrilled when Allen signed her before any other cast member. "My character was a biophysicist, a well-educated family woman who worked in the space program with her husband, who was an astronaut. It all seemed very legitimate and real to me." She didn't balk when the scripts had her playing housewife "sight gags," the mother who, with a push of a button, rolled forth perfectly prepared dinners from her spaceship kitchen, the mother who fed dirty clothes into the top of a contraption and collected them, clean, folded, and wrapped in cellophane, as they dropped out of the bottom seconds later. Because she was a proudly "professional" actor, Lockhart explained, she didn't make a fuss when it soon became clear the mother was going to be peripheral to the family saga. Paul Zistupnevich remembered his friend June Lockhart telling him, "Half the time all I do is say 'Oh, Penny, come home!' or, 'Will! Where are you?' " That did not leave much for the mother of the future to do that would grab the imagination of a boy like me.

After a while the show abandoned its original theme, "all

that early stuff," as Lockhart remembered, "of trying to adapt the home and make sense of civilization out there." After a while Irwin Allen believed he'd found a more compelling device in the tension between little Will, headstrong but resourceful, and the aging scoundrel, Dr. Smith. As played by Jonathan Harris, Dr. Smith was always trying to sell out the Robinsons to an alien if it might mean a getaway for himself. He was spineless and he was clumsy with technology and he was a crybaby whose typical response to a jam of his own making was to snivel, "Oh, woe is me."

The object of contempt at the center or *Lost in Space*, the ever-lurking threat to the family of the future, was weakness and lack of discipline, the whimpering man, the mincing man, the man who refused to be a Man. Was Dr. Smith coded gay? I asked Lockhart. "You said it, I didn't," she answered with a smile in her voice. "You watch the show now," she added, "and you see they put him in drag, they did everything they could. But the mother and father weren't allowed to touch! You know?"

I remember, as a ten-year-old boy, feeling disgust and revulsion for Dr. Smith whenever he was on the screen. I remember wanting Will to just slap him. Was the idea, I asked June Lockhart, that, for the family of the future to survive in space, they had to exert extreme self-control, be ultrarational?

"I think so. The irrationality was taken care of by the character of Dr. Smith."

There were visits to the Gianninis', and there were glimpses of other strange worlds as well, other planets beyond that of my own family's own. There was the home of two neighbor boys, an eerily quiet house but for the ticking of grandfather clocks, a house full of Old West antiques and, taking up much of the family room, a supremely realistic train set constructed by and for the father, who was a radar engineer. There was the other home at the end of the cul-de-sac, across from the Gianninis', a house without children. The man was an older Lockheed worker of rank lower than my father, I knew, and he did not want children on his lawn or anywhere near his house, even when a tennis ball landed there. His wife smiled at us and expected smiles back as if her husband was not an ogre. Once a year she threw a birthday party for her two poodles, inviting all the children on the cul-de-sac to come inside her garage and perform little song-and-dance acts and then to eat cake and ice cream off a long plywood tabletop her husband, nowhere to be seen, had assembled for the event.

There was a planet I found most unnerving, the world inhabited by my school chum John O'Meara, who lived in a slightly cheaper subdivision next to mine. Like me, he was the oldest child (of six) and like me he was an altar boy and a boy who liked to wrestle in the grass. His mother was a sweet woman who always seemed glad to see me, but her house was dark and cluttered with piles of laundry and always smelled powerfully of urine because several of her children wet the bed every night. John's backyard seemed mean, too much cracked cement and clothesline, so going to his house usually meant spending a lot of time in a vacant lot nearby looking for old car parts in the weeds.

I remember the O'Mearas' living room coffee table, though, for the books that lay jumbled upon it, strange literature that John said his father brought home. One paperback was full of terrible prognostications by a man named Criswell. Someday soon people would suffer from "Automatic Atomic Disease" and their bowels would open up and pour onto the ground. Someday soon people would eat frozen human flesh like Popsicles, Criswell said, his crazy ideas accompanied by gruesome pen-and-ink illustrations. There was something wrong about paying good money for a trashy book like this, I knew. And in the same stack I would find more reading that would never be found in my house, a book that straightforwardly illustrated the facts of sex. There were naked people coupling in a book on John O'Meara's coffee table, a book his father *encouraged* him to read, John said, a book I was *free* to look at while Mrs. O'Meara made us peanut butter sandwiches in the kitchen.

From the day I met him, I did not know what to make of Mr. O'Meara, a burly man with an unmistakable dent in the upper-left-hand corner of his high, bald forehead. He took me once, with John and other O'Meara children, to an ear-blasting car race at Laguna Seca and I enjoyed it although I sensed my own father would consider the spectacle somehow debasing. Other than that, John and I did not see each other on weekends; his father, John would say, wanted him. Finally one Saturday when I was maybe eleven or twelve, I had grown restless on my planet and called John up and invited myself over to his. John met me at the door and suggested we get on our bikes and go, but then we heard his father call his name from the backyard. Mr. O'Meara was sitting at a patio table, a small, white round one with tiny rusty cracks in it, an umbrella opened overhead to make a bit of shade in the heat. A couple of neighbor men were there with him, a lot of aluminum beer cans spread before them.

"C'mere, I told ya. Right now!" Mr. O'Meara barked, and I thought that John showed an odd lack of haste in stepping through the dark house and passing through the sliding screen door and presenting himself to his father. There were no more words, just a quick punch from Mr. O'Meara that split John's lip, the lip gushing blood immediately, John turning away with tears but no crying. John met me where I'd drifted after him, a spot at the dim center of the house, and he turned me back toward the door. "Fuckin' drunken asshole," he said. "You'd better go."

I remember looking down the hallway toward the bedrooms and seeing Mrs. O'Meara throwing some more sheets onto the pile of sheets that made her house a throat-gagging house no matter how often she washed the latest pile. John was finding a paper towel for his lip and Mr. O'Meara had turned his big dented skull back toward his friends, and there were dirty dishes on the drainboard as always at the O'Mearas', and there were jumbles of paper and weird books on every other horizontal surface as always at the O'Mearas'. I remember sensing, then, the presence of the monster my parents must have most feared, the encroaching Mess in life that would doom a family, would wreck its controls if vigilance was not maintained. For a fleeting moment, I understood the daily struggle to pass the sweeper, to fold the diapers, to keep a green lawn, to fix every appliance at one's own workbench, to denounce the crass "hucksters" on TV and their dangerous appeals to our weaker natures. I understood why my parents often treated a small thing as if it were a crack in the space hull, as if we were in imminent danger of crashing.

After that Saturday at the O'Mearas', I never visited John's planet again.

Irwin Allen "was an orphan at the age of, I think, ten or twelve and he was raised by his cousin's parents and so he never knew what it was like to have family ties," according to Paul Zistupnevich. He cited this as the cause for why *Lost in Space,* in his opinion, fell short of its potential as family fare. "I've always said Irwin didn't know what a family relationship was. That's why in all his things you never saw a tremendous love story. He didn't know what love was all about. He really was afraid of love. He was afraid of family relationships or expressing love. It was all due to the fact that he was an orphan."

"The reason *Lost in Space* was so successful," Allen's former right-hand man continued, "is that it centered *around* the family. But our biggest problem is that we had too many monsters infiltrate into the thing. That was because of Irwin Allen. Irwin Allen's span of attention was not too great. If the interest waned in the show, you either blew someone up or put them on fire or put them in jeopardy. That was basically his formula. Excitement. He didn't care if it made any sense, he just threw it in."

On the set of *Lost in Space,* as Zistupnevich recalled, other forces threatened the integrity of the Robinson family. "Well, I won't use names but we had an assistant director who wanted

Penny to be a little sex bomb. Oh God, he was always trying. He just thought she should be more curvaceous and of course the problem was that she was supposed to be the young juvenile interest of the thing and so we had to, instead of make her look sexy, play her down. At the time we started out she was about twelve and by the time we finished, in her teens. This assistant director wanted to have her neckline emphasized, the bust and whatnot. We used to have to hit him over the head! And poor little Penny, she felt a little uncomfortable."

In the end, what killed *Lost in Space* was the Mess of life sneaking up and pulling its plug. Ratings were good and CBS was set to renew, as June Lockhart remembered, but when it came time to propose the outlines for the fourth year's string of episodes, Irwin Allen's flighty ego and short attention span prevented him from coming up with any. "All right, just drop it," the powers in New York decided, according to June Lockhart, and the Robinson family was left to die in space the year I turned twelve.

Mother must never be in jeopardy.

Mother and Father must not touch.

When interest wanes, blow up something.

A strange subtext for the favorite program of a boy nearing his teens as the second half of the sixties built up revolutionary (sexual and political) steam. The wonder to me is not that a man like Irwin Allen would think it up, but that I, who was anything but an unloved orphan, would eat it up. Even in the last year, when I felt too old for such childishness, I guiltily watched every episode.

My generation has made a science of revisiting the TV shows we watched as kids—of "deconstructing" them, as the grown-up academics say. Postmodern critics hold that these shows can never be taken at silly face value because they were a welter of messages onto which each watcher overlaid his or her own meanings. A University of Wisconsin feminist scholar named Lynn Spigel, for example, makes the argument that the "fantastic sitcoms" of the era, from *Bewitched* to *I Dream of Jeannie* to *The Addams Family*, can be read as subversive challenges to the ideal of the nuclear family in the suburb. Samantha was the one with the brains and the powers; Jeannie lived unmarried with the astronaut; the Addams Family was an extended family who unmasked the neighbors' loathing of the "other." I find Spigel convincing and I like to imagine a frustrated suburban woman or girl deriving a private solace, even a radicalization, from such patently low-brow television. Political gold, spun from "tripe."

What then, I wonder, did other children look to find in the Robinsons' projection of family? I know that a recent *Lost in Space* convention in Boston drew over 30,000 fans, many of them approaching middle age, like me. I know, too, that June Lockhart, who often visits NASA's Kennedy–Johnson Space Center, told me that "nearly everyone I meet there, to a man or a woman, whether it's a physicist, an engineer, an astronaut, or the guy who zips up the space suits, they all say watching *Lost in Space* made them know what they wanted to do when they grew up."

I still only hazily understand what I wanted to make of *Lost in Space* as a boy. I would guess that I found reassurance in the crisply defined roles within the Robinson clan, the notion that the family of tomorrow was but an emotionally streamlined version of my family of today. The boy *did* exist at the center of the universe, the father *did* provide security through mastery of technology. I would guess that the CBS honchos and Irwin Allen knew me well, knew that some part of me would have preferred a mother who did not touch the father, who worked in rational concert with manly thinking and did not own the power of sexual sway over her man. I would guess that a mother whose technology allowed her to maintain a perfect tract home world appealed to me very much. June Lockhart in her silver jumpsuit, June Lockhart always lovingly *there* on the periphery of the boy's adventure, probably offered me the permission I wanted, permission to close myself to my mother who was loving and who sometimes moped and screamed, who was therefore "irrational," who demanded that I not take her for granted.

Nicky Giannini moved away just about the time that *Lost in Space* went off the air. His father had been transferred by the phone company to a valley brighter, wider, flatter, emptier than our own. They were moving to Fresno in the San Joaquin. By then everyone on the cul-de-sac already knew that Nicky's father had a kind of cancer; you could see it on his neck.

A year or so after the move, Mr. Giannini died and not long after that I visited Nicky in his new neighborhood, my parents dropping me off on their way to a Disneyland vacation for the rest of the family. Nicky's new house, a stucco rancher, looked from the outside much like the one he'd left behind on our cul-de-sac. The interior, however, lacked the precision of old, lacked the tomato sauce smells, too. The hi-fi sat on the carpeted floor of the living room against a bare wall. You had to squat on your haunches to put on a record and there were only rock 'n' roll records, no Dean Martin, no Jimmy Durante. Nicky's mother was not feeling well during the week I visited, spending most of her time in the bedroom with the door closed.

Celeste had remained in Northern California where she was living with a hippie artist. Joe, working two restaurant jobs in Fresno, rarely materialized around the house. The yards front and back were neat but lacked the elaborate beds and shrubs that Mr. Giannini had insisted on when he was alive on our cul-de-sac. Beyond the cinderblock wall that ended Nicky's new backyard, a field of baked clay stretched for blocks until it reached another low line of cinderblock. The Fresno heat sucked a person's energy, made the insects snap and buzz angrily. Still, for something to do, Nicky and I rode our ten-speed bicycles beyond his neighborhood, beyond the edge of Fresno, into the first foothills of the Sierra Nevada, sweating and gasping and pumping our way up to the reservoir behind Friant Dam. Looked upon from that height, Fresno shimmered in its valley like a flattened can on a dirt road, a bit of metallic sparkle but nothing worthy of a person's imagination. On the ride back, farmland gave way to strip malls which gave way to broad boulevards cutting through subdivisions until finally, throats burning, thighs aching, we turned a corner and Nicky's air-conditioned house swung into view and I saw it as Mr. Giannini and Mrs. Giannini must have wanted to see it when they moved in: as a private oasis rather than as an idea of self-enclosed family life taken too far, taken to the edge of nowhere.

Not long after, Mrs. Giannini died unexpectedly. I heard my parents discussing it with some of our neighbors, nodding their heads about what I had never been told, that Mrs. Giannini, for as long as they'd known her, had been given to bouts of depression. A mixing of medication and alcohol was said to have been involved, though there was no way of knowing if Mrs. Giannini had taken her life intentionally. More likely, everyone agreed, a sad accident was to blame. Mrs. Giannini had been a good woman, a good mother, everyone said. But she could not, in the end, be preserved from the jeopardy posed by her melancholy. She died while napping behind the shut door of her bedroom.

SIX

CHLORINE

My young adolescence corresponded with that time when America's ghettos burned, when the Vietnam War escalated, when campuses roiled in protest, when older children of the suburbs lit out for hippie hedonism. I spent much of that time immersed in water chlorinated and filtered clear, water heated at the proper temperature to be bracing but not chilling, water that smothered sound and turned the black line on the pool's bottom into a blurry focus of grunting meditation.

My parents had decided that life in our cul-de-sac world did not offer the proper structure for a boy my age; I was finding it too easy to waste time. I was watching too much television, reading too much *Sports Illustrated,* spending too many afternoons playing Battleship and Poker and Monopoly with other boys on our backyard deck. Following the lead of friends who lived one cul-de-sac over, my parents had joined a swim and racquet club and they had placed me on the swim team. "Just try it and see how you do," my mother had said.

I was the skinniest eleven-year-old I knew. I would stare with dismay at my shirtless chest in the bathroom mirror, my ribs in such clear relief, the bony points atop each shoulder that no other children seemed to have. I was the skinniest boy on the Kona Kai swim team and no natural as a swimmer. "Pick it up, Hat Rack!" the coach would shout at me as I reached the end of the pool and let the fast, experienced kids by, their legs gracefully flopping over and twisting off the wall in a perfect flip turn, a maneuver that confounded me for the first many months of swim practice. "Hep! Hep! Hep! Hep! Hep! Hep!" the coach, a former military man, would chant in rhythm to the rolled-head breathing of fifty children churning the water with crawl stroke after crawl stroke.

An afternoon's workout would be about one hundred and eighty laps of the twenty-five-yard pool. The total (about two and a half miles) would be divided into distances and strokes and paces of the coach's choice, which he would write on a chalkboard for all swimmers to see: *200 Warm-up. 10 × 100 freestyle on the 2 minutes. 5 × 200 IM on the 5 minutes,* and so on. This made the true master of the workout a big clock that stood at one end of the pool, its foot-long hands—red for seconds, black for minutes—pointing silent accusation at the swimmers who fell behind the pace, who finished one distance only to look up and see they were already late for starting the next. The slower swimmer tended to get less rest, less time with head above water, and so a workout with fifty children felt to me a very private struggle. My eyes, smarting from the chemicals, would study the black line below me the way a prisoner memorizes ceiling cracks over his bunk. My skinny chest would beg for air after the first twenty strokes; by the last twenty the muscles in my back and butt and upper arms would ache as if I'd been pummeled. After workout I'd race with the other boys to the steamy warmth of the locker-room shower. That was the only pleasure of any workout as far as I was concerned, and yet the children around me all seemed so content to have been placed by their parents on a swim team. They had just tried it to see how they would do. They seemed to be doing quite well. They did not complain of boredom or sore muscles. Indeed, there seemed a quality of feral happiness about the boys who jackknifed into the pool when workout started and snapped each other with towels in the locker room afterward. They were smooth-bodied children without bony points on their shoulder tops and I wanted what they had, that body, that feral happiness, which, I assumed, would come to me once I had stroked enough thousands of armstrokes.

My parents were pleased that I decided to stick with the swimming, making myself stiff and tired every day following school and, during the summer, twice a day with workouts in the morning and afternoon. My mother did not mind the time she spent driving me (and eventually my brother and sisters) back and forth to swim workouts. "What I like about swim team," my mother liked to say, "is the kids come home clean."

Those were years when my father grew more pensive and grouchy, as if shaken awake from a dream he could never get back to. As his volatility increased, he required all the more surveillance, and I, having trained myself well, often saw one of his explosions coming while everyone else in the house ignored or even unwittingly fed his frustration. My mother could set him off just by yelling from one end of the house, "Hal? Have you seen the checkbook?" Simply by making a demand on his attention she would have invited his grouchiness upon her, and soon the small matter of a misplaced checkbook would be transformed into the much bigger matter of who had been the first to bitch at whom, who was *complaining* for *no reason*. One afternoon in the middle of an argument with my mother, my father, who had always been a shouter when he lost his temper, did an uncharacteristic thing. He turned and punched the wall in the entryway of our home. To my astonishment, his fist made a hole through the drywall, a hole my father stalked away from without any further comment. The hole remained there until the next morning when my father came to it with his toolbox and a can of paint. He carefully made a patch, but there was a faint outline if you knew where to look.

Sometime around the beginning of 1968, a time when the slaughter of the Tet offensive was playing nightly on television and United States aircraft were carpet bombing North Vietnamese civilians with the official aim of inflicting "punishment" and "destroying morale," my father decided that he was going to build an airplane of his own in our garage. He purchased a box of blueprints for a Thorp T-18, a "home-built" two-seater that was pictured airborne (pilot and passenger clearly smiling) in a magazine he showed me. He bought a four-cylinder Lycoming engine that needed some rebuilding and stored it in a corner of the garage. He bought some new tools: a Whitney hole punch, a pop rivet gun, tin snips, aviation scales, precision rulers. He came home with a huge slab of particleboard sticking out of the back of the Dodge 440 and from this he constructed a ten-foot-long worktable with retractable legs and eye-hooks on the corners. With a rope and pulley, my father would winch the table up to the ceiling of the garage where it hung in storage above the two cars until the next time he was ready to work on his airplane.

For a while, my father spent evenings and much of the weekends in the garage making small parts for the Thorp T-18. Some of them looked very promising, airfoil-shaped pieces of aluminum my father had carefully cut and bent and shaped around plywood templates. Whenever he finished a piece he hung it on a hook over his workbench. But after several months the hook held only half a dozen of the many hundreds of pieces needed to make up a Thorp T-18. My father came to the conclusion that he had neither the time, money, nor room to build a do-it-yourself airplane and so he sold the plans, the engine, the table, and most of the tools to another man who came one Saturday and collected everything in a pickup truck.

After that my father moved his hours of quiet aloneness inside of the house to the living room couch. A friend at work had piqued his interest in books by thinkers, books that tried to make sense of society and human nature and where the country was headed. My father started carrying these thick books with him to and from work. He would read at Lockheed on his lunch hour and in the living room after dinner. He became a student of John Kenneth Galbraith's critique of corporate life as the misplaced pursuit of money over life's other, more fulfilling rewards. He soaked up Eric Hoffer's idea that mass movements are the products of adults gripped by "juvenile" restlessness, which, in turn, is produced by economic dislocation due to technology. He became engrossed by the field of general semantics as explicated by Stuart Chase, who warned that meaning in language is nothing concrete, that people too easily fall for propagandists who pretend "the word is the thing." He read Desmond Morris's theories about humans as "naked apes" gone neurotic in the too-crowded cages of the inner cities. He read a number of books about race and prejudice, including *The Autobiography of Malcolm X* and Eldridge Cleaver's *Soul on Ice*. He took to heart the portrait of a shrinking planet offered by Paul Ehrlich and Alvin Toffler. My father's titles ran into one another like haikus of gloom . . .

The New Industrial State (Galbraith)
The Human Zoo (Morris)
The True Believer (Hoffer)
The Temper of Our Time (Hoffer)
The Tyranny of Words (Chase)
Crisis in Black and White (various contributors)
The Population Bomb (Ehrlich)
Future Shock (Toffler)

. . . a brooder's bibliography. These were thinkers suspicious of group-think and the pronouncements of officialdom. They were voicing deep dissatisfaction with how things were and, as well, equating maturity with the sober acceptance that limits must be obeyed, that affluent America must recognize there could be too much of a good thing. The serious grown-up, these books seemed to say, questions every premise of blue sky optimism.

To my grateful surprise, my father's books formed a new bridge between us. He would be well into his latest and want to share with someone his head full of ideas. If he had me there beside him holding the flashlight for a fix-it job, he might say to me, "I'm reading an interesting book, Dave," and then, in the wholly adult language he always used with any child over the age of four, he would begin laying out the book's thesis and making connections to his own experience. He would speak of an insight he'd gleaned about Lockheed: "The primary purpose of any bureaucratic institution is to continually justify its own existence." He would be excited by a semanticist's version of why a husband and wife talk past each other: "Words are useless unless two people agree on their meanings. The map must agree with the territory!" He would toss such wordy nuggets at me as if I were his peer and not a boy of twelve or thirteen who mostly read novels about sea adventures. Though I obviously grasped little of what my father tried to convey, I found it exhilarating that he had thrown open this door to his mind and had invited me inside. He seemed to be asking me, as a potential equal, to accompany him on an intellectual journey just begun.

The restrained, rationalist tone of the economists, the semanticists, the social scientists won my father's trust. "Here's my test of any author's hypothesis," my father would tell me often during our impromptu discussions. "I look around me and see if my experience matches the theory put forward. In the case of Galbraith" (or Chase or Morris or Hoffer or whoever) "I find his hypotheses to be borne out in the real world."

In his books my father found schematic diagrams that might offer solutions to a malfunctioning America or, perhaps, to the uneasiness in his soul. He was the troubled troubleshooter, a systems engineer in search of systems underlying the human condition.

My mother's frame of reference on those times was quite a different one. She did not hunger for systemic explanations. She remained secure in her faith in the one system revealed through the Catholic Church. The wars and the race riots, all the day's disturbing news could be read as evidence of the continuous struggle between God's grace and man's sinful nature. Clearly there was not yet enough human goodness in the world, for which failure every person shared some measure of collective guilt. In this she reflected the thinking of a Catholic she had read and admired in college, the Trappist Thomas Merton, who told himself at the dawn of the Second World War: "I myself am responsible for this. My sins have done this. Hitler is not the only one who has started this war: I have my share in it, too . . ." The more hellish the headlines, the more the Catholic was called upon to pray and confess and lead a virtuous existence. And, too, to scan the horizon for saintly examples who might offer cause for hope.

My mother found embodiments of hope in two Catholic men: Cesar Chavez and Bobby Kennedy. Chavez, who spoke of St. Francis as his inspiration and who enjoyed the support of Catholic bishops, was asking suburban mothers to boycott California grapes, and so my mother did. Cesar was "for the poor" (as all saintly Catholics must be). Bobby, too, was "for the poor"; more so, in my mother's view, than any other presidential candidate. On the June morning in 1968 when Bobby Kennedy died of shots to the head, I woke to the sound of the television at an hour still black outside. In the family room I found my mother alone under a blanket on the yellow Naugahyde couch, devastated by the martyrdom replaying every few minutes on the screen.

My mother was a Kennedy Democrat with unreserved sympathy toward the struggles of blacks and Mexican grape pickers in the United States, but it was to a different cause, a different saintly endeavor, that she devoted her charitable energies during the late 1960s. Queen of Apostles parish had adopted as its missionary an American woman named Mary Nouveau, who lived in a high desert region of Sonora, Mexico, with the Tarahumara Indians. Their brown faces hung in the church foyer, telling me that through the Catholic Church I was connected with their otherwise unfathomable lives. Every Sunday my mother collected ten dozen doughnuts from a shop and sold them after Mass at a folding table just outside the church doors. All the profits went to Mary Nouveau and the brown faces, helping to increase in some small way the level of human goodness in the world.

Here, then, was the odd truth about life in a blue sky subdivision at that time. The social upheaval occurring fifty minutes up the road in Berkeley and San Francisco felt further removed from our existence than the cave dwellings of the Tarahumara Indians. I remember a rare family outing to San Francisco during those years, a Sunday spent on the green and tidy side of the city at the Palace of Fine Arts with its lagoon covered by swans and ducks. When we returned to the Dodge 440 and unlocked its doors for the trip home, a strange stink of burning met our noses. "What is that?!" we all said as we settled into our places in the station wagon.

"I wonder," my mother said, "if some hippies might have broken into our car and smoked some marijuana in here."

After a few sniffs, my father decided the actual source of the smell was a shorted-out wire in the car's electric-powered rear window.

More than a million armstrokes through water, three years of workouts, proved me to be a quite average swimmer in all events but one, the fifty-yard breaststroke, at which I was slightly above average. This was official because the Amateur Athletic Union, based on the times of every registered swimmer, had established a curve of performance broken into "B" level (the majority of swimmers including me), "A" (better swimmers), and "AA" (potential Olympians). The only "A" time remotely within my reach was the breaststroke standard, which is what my father wanted to discuss with me one afternoon in the living room.

He had been doing some figuring in the small spiralbound notebook he always carried in his pocket. By his calculations, the few seconds separating me from "A" time need not be as daunting as I made them out to be. How many strokes, my father asked me, do you take in a fifty-yard race? I made a guess. He folded that figure into his equation and solved. To be an "A" breaststroker I need merely pull myself an extra three inches farther per stroke. "Three inches," my father said to me. "If you think about it, that's about the length of your fingers outstretched. Work extra hard at practice. In a few weeks, at the championships, make up just that little bit more distance with each stroke, and 'A' time is yours."

Who was I to argue with the numbers in my father's notebook?

The swim meet was a communal ceremony for a corporate tribe. Unlike Little League, which gave Mother and Father nothing to do but cheer behind a fence, the swim meet could not go forward without layer upon layer of parental involvement. For every lane in the pool there needed to be three parents clicking stopwatches and another who wrote down each swimmer's name, time, and order of finish. For every heat swum, a parent was needed to bring the records from the pool to a central point where other parents tabulated the data and posted official results on the proper forms. A parent was needed to declare "Swimmers take your marks" and to shoot off the starter's pistol, another to make announcements over the loudspeaker. Parents were needed to write names on every prize handed out, to set up all the tables and chairs early in the morning, to break everything down after, to sell doughnuts and coffee and hot dogs during and then to account for the proceeds.

The swim meet was a tribe's beliefs manifested for its children. Belief in the power of tasks highly organized (a role for all). Belief in performance measured precisely (each child's progress monitored to the tenth of a second). Belief in hierarchy (each child ranked in exact relation to every other child). Belief in meritocracy and its rewards (ribbons for places six through four and medals for three through one). Belief that repetitious rigor in a sterilized environment was best for children ("they come home clean"). This is why competitive swimming flourished in the affluent aerospace suburbs of the Sunbelt during the sixties and early seventies. The practice pool of El Monte Aquatics, the powerhouse of Southern California, lay within twenty miles of Cal Tech, the Jet Propulsion Laboratory, and plants of Northrop, Rockwell, Hughes, and Lockheed. In Northern California, the families of Lockheed and Ford Aerospace and GTE and other military contractors supplied swimmers to two world-class teams: De Anza Swim Club (650 members, and Santa Clara Swim Club, with its perennial national champions). In such places, for every superteam there were hundreds of little clubs like Kona Kai, hundreds of swim meets underway every weekend during the summer months.

Meets that pitted one team against another lasted half a day, while much larger ones, open invitationals for "B" swimmers, filled an entire weekend and drew many hundreds of participants. My thoroughly involved parents experienced the meets as busy blurs, but for me they were something else again. For me they were not really about the manic plunges into cold water that lasted a minute or two at the most. They were about all the time passed in between the day's races.

Swim meets for me were about lazy yearning, hours spent lounging with other young teenagers on sleeping bags made warm by the sun above and the concrete beneath. Naked but for a thin film of Speedo nylon, we lay and played cards and sucked honey out of plastic bottles shaped like cute bears. I would lie there and want. I wanted the beautiful girls whose nipples showed through their wet suits, whose pubic hair, incredibly, peeked out from the elastic around their vulvas. I would sit this way or that to hide my erections, the Speedo useless as disguise. I would lie and look and craft in my head the funny thing I would say to one of the beautiful girls, though usually I did not say it or if I did, she would give me the quizzical look I knew I deserved. She and I would know that within the seemingly random sprawl of young bodies on sleeping bags there existed a social order. I was skinny with freckles and zits, a "B" swimmer. The beautiful children (smoothly muscled, smoothly tanned, smoothly featured), they ruled the drowsy interludes between races with yawns and stretches and decisions about who to play cards with, whose sleeping bag would be spread next to whose. They tended, as well, to be the "A" swimmers.

I did not reach "A" time in the championships that summer, although I did take three tenths of a second off my time for the fifty-yard breaststroke. I never did, in fact, achieve an "A" time.

My father's company made one of the jet fighters that strafed and bombed North Vietnam, the F-104. Lockheed also built the airplanes that spied on the Vietcong: the supersonic SR-71 Blackbird, the U-2 jet, and a secret airplane called the QT-2, a converted glider fitted with a muffled engine, a wooden propeller, and infrared sensors. The QT-2 would quietly fly low over the jungle at night, pinpointing enemy movement so that the B-52 pilots knew where to drop their tons of bombs the next day. Reconnaissance pictures of the Ho Chi Minh trail were snapped by Lockheed's CORONA satellite. And it was my father's employer that built the fat-bellied transports that brought weapons and ammunition and hundreds of thousands of young Americans in and out of the war. It seemed you could not watch a television news report about Vietnam without seeing turbo-prop C-130s or jet-powered C-141s and C-5s taxiing in the background, Lockheed products all.

At the time, in an effort to rise within Lockheed Missiles and Space Company, my father pursued a master's degree in mechanical engineering. He would study at a card table in my parents' bedroom under a window that looked out on the front lawn. On the other side of the window, the children of the cul-de-sac would be machine-gunning each other, little bodies flying everywhere. My father would stick his head out and yell, "War is not a game! Knock it off!" But his own thoughts were not so easily chastened. Once he had his master's degree, after he had decided he could not build a real airplane in his garage, he began spending hours at his workbench meticulously assembling and painting plastic models of the warplanes of his youth. They included the heroic Spitfires and Hurricanes that won the Battle of Britain, and I remember my father's eager anticipation the day he took me to see the new movie about that air battle, a film heralded for its realistic dog fight scenes. I remember that when a German gunner got it, his goggles filling with blood, the audience cheered while my father only stiffened, clucking his tongue. When my father's models covered the top of the television and a bookshelf, he gave a few to me to hang over my bed. He also, some of those autumns, took me to the Reno air races, where for two days we would watch the Bearcats and the Mustangs—World War II killing machines now painted like giant toys—chase one another around the desert.

My father was fun to be with at the Reno air races, fun to be with, too, on the several backpacking trips we made in those years. We went with another Lockheed father and his two boys, hiking all day in Desolation Valley north of Lake Tahoe. "We don't dilly dally in Desolation Valley!" my father liked to joke as we made our way through the granite and pines. When we reached our lake at the end of the day, my father baited my fishing hook with salmon eggs from a jar labeled "Potsky's Balls o' Fire." "Ready or notskys, here come Potsky's!" he shouted as I cast my line. The other Lockheed boys had that feral happiness I recognized in my swim teammates. They hiked a dozen steep miles without showing any strain and upon arrival caught trout with their bare hands in a stream behind camp. I did not feel much in common with them, but it did not matter because my father and I could walk together and talk about the books he had been reading and how great it was to be in such a setting so magnificent and ancient as the High Sierra.

On the drives home, whether from the Reno air races or a backpacking trip, my father would seem for long stretches of road to have run out of topics and words, though at some point he always offered me the same advice. Working for a big corporation meant a life of compromise, he would say. I should avoid taking on too many obligations too soon, he would add.

By the summer I was to turn fourteen, I could no longer doubt that the children of feral happiness would always be different from me. Now they were growing ripply muscles in their thighs and chest and arms. Their backs were ever broadening V's and their blond hair, tinged an enviable green by chlorine and sun, curled with fussless perfection on the napes of their necks. Some of these kids had grown this way while swimming half the laps I did. And yet there I am in a team photo, my chest still a lattice of bone, my arms crossed with a fist wedged behind each bicep in order to spread the flesh and create the illusion of *some* muscle. At the annual team slumber party, the beautiful boys and girls slipped in and out of each other's sleeping bags while I snickered with other "B" boys. When the championships rolled around, some "A" swimmers who had quit the team were invited back in order to help Kona Kai get extra points. With little more than a few days' working out, they swam their "A" times and walked off with medals ahead of me.

Swim team was teaching me, by then, lessons the tribe had not intended. Hard work did not automatically ratchet one to the top of the meritocracy. Rather, the gifted lived there always, and accepted their places with the equanimity of natural aristocrats. My brother, six years younger than me, who had been given the body I wanted, joined the swim team at age six and won virtually every event he attempted over the next three summers before quitting and going on to be a tennis champion. I saw this, I saw the ever more quizzical looks the beautiful girls gave me, and I made my refuge in sarcasm.

Quickness with words, an instinct for the ironic aside, was my only natural talent, I was beginning to understand. Words felt comfortable in my head, before my eyes, on my tongue, in a way that water never had felt comfortable around my body. My problem was that in the swimming pool, the gift of glib held no currency. With every reach and pull through the water, words came and went, came and went, slipping through my mind, rushing out my nose and mouth in clouds of bubbles. But my teammates, my coach, the clock at the end of the lane, could not care less for my words. The swimming pool was meant to be a zone of raw, blind exertion; that was the cleanliness of it. "Here's a dime, Hat Rack. Phone someone who cares," the coach would say to me whenever I'd try to kid him into changing the terms of the workout. What he saw in me, what I was becoming despite my thousands of laps in the swimming pool, was one more skinny, smart-assed teenager. Smart-assedness, in those days, permeated the air like freshly released pollen: smart-ass signs waving in protest (Vietnam: Love It or Leave It), smart-ass stickers stuck on bumpers: (Burn Pot, Not People), a smart-ass name for our President (Tricky Dick). It was as if the country had just discovered the teenager's tool of subversion, a tool wielded by me with improving confidence.

I had powerful incentive. I was learning how, with the right crack at the right moment, to disarm my father in one of his testy moods. Now when I saw an explosion coming I might move beyond mere surveillance to intervene, might manage my father's emotions before they could go critical. I could observe his eyebrows gathering and I could say some words that put the eyebrows back where they belonged. My father's appetite for cynical humor seemed to grow with every book he read, a craving I could satisfy if I chose my words, cuttingly funny words, carefully. If I didn't, of course, the risks were high. What if, in one of his moods, my father were to sense that I was making him the *object* of my smart-assed remark?

I do not with any vividness remember the sight of Neil Armstrong setting foot on the moon. I know that at 7:56 on the evening of July 20, 1969, my family was together in the family room, most of us crowded onto the yellow Naugahyde couch. But the history-making we saw was a television image, and so it has become, after countless replays, a visual icon that blots out any memory of that first perception. What I remember is the conversation around the event. My father had known Neil Armstrong as a classmate in the aeronautical engineering program at Purdue University. This fact seemed to me a kind of benediction upon my family, and I asked my father all about Neil Armstrong. All my father could recall was that the first man on the moon had been a friendly fellow, low key, happy to drink a beer with my father.

I remember my father's reaction to his college mate's now-famous phrase: "That's one small step for man, one giant leap for mankind."

"You don't say a thing like that spontaneously," my father said. "Neil got some help. At minimum, he spent weeks working out that line."

I remember, too, my father saying the Apollo program was an unnecessary risk of human life, a "dog and pony show for the taxpayers." Satellites, he told me, could do anything a person could do in space and "faster and cheaper." Indeed, Surveyor I had made a perfect landing on the moon more than three years before, yet America reserved its exultation for Neil Armstrong's footfall.

I remember my father the next day pointing to an article in the newspaper about people who preferred to believe the moon landing had been faked. "Not that farfetched, really," my father said. "The political futures of powerful men are riding on this. They've promised the nation it could be done and what if it couldn't? Would they just shrug and say, 'Sorry about that, no can do'? Hardly. All you'd need is a big television studio and a few dozen souls willing to go along with the charade for the sake of national security. Not that farfetched at all."

I remember a moment some weeks later that has become the stuff of legend in our family. Neil Armstrong was back on Earth and out of quarantine. He was in Washington, D.C., making a televised address to politicians. My father was in one of our bathrooms occupied with a fix-it job. "Hal," my mother called to the back of the house. "Neil Armstrong is giving his speech."

"Not now, Terry," my father called back. "I'm deep into this." And then a pause, and then a wail from my father that was only mostly in jest. "My God!" he said. "How our lives have diverged! Neil Armstrong is being applauded by the President, and I'm staring at the underside of a toilet!"

"This," President Nixon declared of the lunar landing, "is the greatest week since the beginning of the world, the creation." My tribe, certainly, assumed Apollo's success would secure our dominance. Having been placed in charge of the future, we had delivered as promised a man on the moon on television. As scripted by our tribe's father, Wernher von Braun, the moon shot was to be but a way station on the path to Mars, a warm-up for the building and launching of a "flotilla" of manned spaceships to the red planet.

Yet almost at the moment *Eagle* landed, blue sky icons seemed to lose their mojo powers. Environmentalists, utterly unmoved by the sleek loveliness of the SST, painted it loud and rude, shot it down. The B-70 superbomber program had already crashed in Congress as the antiwar movement challenged the notion that the Pentagon, NASA, or any state authority should be the revered and unchallenged steward of a people's imagination about the future. Few (except for my tribe) took seriously Vice President Agnew's call for a Mars mission.

Seventeen weeks after the greatest week since the beginning

of the world, one day after a second successful Apollo mission had returned to Earth, Washington, D.C., was engulfed by more than a quarter of a million dissenting Americans, the largest antiwar demonstration in the nation's history. Their slogan was "March Against Death." Citizens across the country were asked to wear black armbands to indicate their support for the moratorium. My father wore a black armband into his place of work at Lockheed Missiles and Space Company. A colleague said to him, laughing, "Hey Hal, who died?"

"About fifty thousand Americans and untold Vietnamese," he answered flatly.

An hour later my father was invited into the office of his supervisor, who informed him that wearing a black armband "around the project" was the sort of thing that could get a man's "ticket pulled." My father knew others whose security clearances had been lifted with less explanation; the giving and taking away of secret privilege was itself a secret process. He knew also that with no ticket he would be without a job at Lockheed, and he needed the job. He removed the black armband and returned to his desk.

One year after the greatest week since the beginning of the world, a slowing in weapons spending, coinciding with a slump in commercial jet sales, caused mass aerospace layoffs across the country. California, where the livelihoods of some half a million people were directly tied to aerospace, was hardest hit. On November 1, 1970, when Richard Nixon came to our city of San Jose to campaign for a Republican senator, the nation saw his car pelted with bricks and bottles, saw the President cursed and denounced by demonstrators against the Vietnam War. What did not make television was a group of protesters wearing suits and ties, carrying picket signs with a different message. They were unemployed aerospace workers like Rudy Rider, a married forty-four-year-old engineer with two daughters. "I wouldn't want to get involved with anything like that," Rider told a newspaper reporter as the antiwar melee broke out across the street. But he was there to remind politely that "We're not statistics. When you're out of a job, you're one hundred percent out of a job." A fellow picketer added, "It's in Nixon's lap. Federal dollars put these people in this county and federal dollars should take care of them."

That was a time when there was a call for a "Cesar Chavez with a Ph.D." to lead aerospace workers in marches on capitol buildings, a time when several dozen unemployed aerospace workers did march on the California state capitol, vowing to nail a list of demands to Governor Reagan's mahogany door if he did not see them. "We just want meaningful jobs which will enable us to live a life consistent with human dignity," read their proclamation. That was a time when some of my father's peers went from designing satellites to perfecting alloy backpack frames (surely Dad hadn't been a compatriot of Neil Armstrong's only to make camping gear!). I remember, while watching the Jetsons suburbanize the Milky Way on TV, reading newspaper stories about out-of-work Lockheed men killing themselves.

That was a time when the tribe clung to hubris nevertheless. One year after the greatest week in the history of the world, members of the blue sky tribe convened the West Coast's First Aerospace Congress to take up the issue of where next to apply the methods used to conquer the moon. Let us now turn aerospace engineers loose, speakers urged, on the problems of pollution, crime, urban blight, racism, poverty. Let us fix society with systems engineering. It fell to Vernon L. Grose, vice president of Tustin Institute of Technology in Santa Barbara, to counsel some caution. Yes, he saw opportunity in "the list of civil problems . . . long and getting longer." But the systems approach works best when human variables are kept to a minimum. "If people could somehow be eliminated from socio-economic problems, the solutions would be quite straightforward," he offered in the paper he delivered. "However, because people play such a dominant role in the problems, optimism becomes reserved."

Eighteen months after the greatest week since the beginning of the world, *Aviation Week and Space Technology* was urging America to get back to the only business that aerospace companies really were cut out to do: the making of weapons and space vehicles. The year 1970, concluded the journal's editors, had been "the gloomiest year in decades" and they knew who to blame: the sorts of people who wore black armbands, barbarians against the tribe. They blamed "the mounting assault on technology by a strange coalition of political opportunists, disgruntled youth, ecologists, and advocates of the social welfare state."

That was a time when my father ended an argument with my mother by flinging a wineglass into the corner of the kitchen where it burst into shards. A time also when my father kicked at a bathroom door, leaving a splintery crack he found impossible to patch. That was a time when my father would sulk all weekend long, increasingly immune to my words, often angered all the more by them, by my "attitude," by my "tone." I remember him one day grabbing me by my T-shirt, lifting me off the ground and against the wall of the house, saying he was sick and tired of my smart-assed remarks, raging into my face as the hot stucco prickled my back.

The summer of 1972, the summer I turned fifteen, was to be my final summer on the swim team. In June, Nixon operatives were caught bugging the Democratic National Committee's headquarters in the Watergate complex. In August, on the day after the United States withdrew its last combat unit from Vietnam, B-52s pounded that land with the most intensive twenty-four-hour bombing raid of the war. I floated through that summer on glints of sunlight and whiffs of chlorine, dreaming of sex with beautiful green-haired girls but having sex only with myself. My duties at home were minimal: keep a neat enough room, keep the grass mowed. Most of the time I swam and played tennis and Ping-Pong. Nixon's sending of American players to China had made Ping-Pong a craze of sorts. There was a Ping-Pong table at Kona Kai where I would play for hours with swim teammates. My father had bought a table for our backyard deck, where my brother and I would slap the plastic ball back and forth endlessly, perfecting our top-spins and slices. At the beginning of the summer my father had been able to defeat me handily, but as the season wore on I did better and better against him.

There came an evening late in August, a hot evening with a sky still light and grasshoppers chirping madly, when my mother was off at some church function and my father was doing the dinner dishes. I cajoled him into coming outside for a game of Ping-Pong, a game that turned out to be the first I'd ever won against him. It started out with laughs from my father, his cheers for me when my put-aways skittered off the corners of the table or bounced against his chest. By the end, when I beat him with a back-handed slam, my father had withdrawn into one of his sulks. He dropped the paddle on the table and went back into the kitchen where he muttered as he loaded the dishwasher.

I felt badly for my father, responsible for his sulk, and so when my brother took his place at the other end of the Ping-Pong table, I loudly said to him, "You know, if Dad got to practice as much as we do, I never would have won against him."

My father's shout came from the kitchen, through the sliding screen door. "The last thing I need, Dave, is for you to *patronize* me!"

He was onto me. He would not, after all, be fooled by my methods of emotional management. He was turning my words against me as evidence of something even worse, proof of bad faith. That panicked me some, angered me, too. "You may *think* I was patronizing you, Dad," I shouted back, "but I wasn't!"

I saw him then striding quickly from the kitchen, his face red and hard, one of his arms (the arm of a six-foot, two-inch man who was not skinny) throwing open the sliding screen door as I backed away. I saw his clenched fist before my eyes and then blackness and sparks on the backs of my eyelids as another fist and then another crashed into my cheeks and eyes. I tumbled back and found myself sitting, ridiculously it seemed, on the soft plastic cushion of a chaise lounge. Through hands I'd thrown over my face too late, I watched the back of my father as he disappeared without a word through the screen door, into the house. I heard my brother whimpering from the hiding spot he'd found beneath the picnic table. I felt my face not hurting so much as pulsating. Except for the whimpering of my brother and the running of my father's dishwater and the singing of the grasshoppers, all in our backyard was quiet and warm and removed from the world as usual.

I walked then, barefoot and wearing cut-off jeans and a T-shirt, out of our backyard, our cul-de-sac, past Queen of Apostles Church and school, past the softball field by the freeway, through the subdivision of homes that looked just like ours except with second stories, under the overpass where Interstate 280 meets Lawrence Expressway, across the AstroTurf lawn of a corner gas station, down a strip of burger stands and car dealerships to the Futurama bowling lanes. Within miles of our house, the Futurama was the only public space I could think of where a teenager could be with people, yet anonymous and alone late into the night. I sat and watched people bowl for an hour or two and then I went into the bathroom to examine my now-throbbing face. In the mirror were blackened eyes and bloody-rimmed nostrils. I knew my face would only look worse the next day, when the blue sky summer was expected to resume as normal.

When the Futurama closed down at midnight, there was nothing to do but return home, to open the front door my father had left unlocked, to move through the dark house to my bedroom, to ease myself under the sheets. I lay there a few minutes and my father came into my room without turning on the light. He sat on the edge of my bed and whispered an apology. There was "no excuse" for what he had done, he said. I was silent. He whispered a theory about the rivalry that occurs between two males in confined space, a theory I knew he'd drawn from one of his books. I was in no mood to accept his apology, much less to theorize with him. I remained silent as he said again, "But for all that, there's no excuse for my actions," and left the room.

The next morning my mother cried, "Oh, no! What did he

do?'' when she saw my face. When next they were alone together, she no doubt demanded an explanation of him. I would guess that he told her what he had told me, and that she forgave him having reached an understanding that he would never do it again. My father never did hit me again.

At swim practice I was grateful for the chance to put my head underwater, to be left alone with newly bitter words as I stared at the blurry black line. At the next swim meet, however, the dive off the blocks and the somersault turns forced cold, chlorinated water up my nose, searing the nasal passages my father had crushed with his fists. The sprint to the finish was all fury and when I looked up to get my time, one of the adults holding a stopwatch was my father, reading the numbers off matter-of-factly.

''What happened to you?'' my teammates kept asking. I'd tell them I had been playing Ping-Pong, had lunged for a put-away but had run my face into the pole that held up the bamboo roof over our deck. My father once heard me telling this story and he took me aside. He told me never to lie on his behalf. ''What I did was wrong. You go ahead and tell anyone you like exactly what I did. Just tell the truth and let me suffer the consequences.'' I listened silently, staring away without forgiveness as my father said this in a voice that was tired, even gentle.

SEVEN

ANOTHER LOCKHEED SON

On a bright morning in the autumn of 1994, I sat at the big oak table in the dining room my parents had added onto their ranch home some years back. The hour was early and the house was quiet. The day before, the children of my sister, Marybeth, had filled the sunny rooms of the house with their shouts and laughter and the electronic beeps of their toys. Now there was only the sound of the coffeemaker gurgling, my father in his pajamas clearing his throat across the table from me, the pages of the Saturday newspaper rustling as we read and waited for the coffee to be ready.

I was in town to conduct interviews, having long ago made words my livelihood. During my years as a journalist I had freelanced for many publications (once, even, *Vogue*) and I had held a few editing jobs, most recently at *Mother Jones* magazine, a consistent basher of the "military-industrial complex" that employed my father. I had found words with political implications the most interesting ones to write, words that strove for the ring of reason and fact but which, in the end, always appealed to the murkier aspects of human nature, moral sensibility and fear and faith.

None of my articles had ever required of me the technical precision, the obedience to immutable natural laws, that engineering had demanded of my father. Journalism had asked merely that I know how to arrange certain pliable arguments and evidence in interesting contraposition to others, and that I maintain a sharp interest in some aspect of the world for a matter of weeks or, at the most, months. Journalism had allowed me to live wherever my wife, Deirdre, and I happened to want to live; at the moment, Vancouver, British Columbia. Though there was little financial security in my sort of work, I continued to like it well enough and it often brought me back to the Bay Area where I could visit with my parents for a few days in the house where I had grown up and where there was always a room with a soft bed and a chest of empty drawers awaiting my arrival. The house and the room were still there for me because, all those years, my father had stayed put at Lockheed, had stuck with aerospace engineering, had with my mother paid off the mortgage. He had followed his original life plan, believing it on the whole a sound one.

Anyone who knew no more than these things about the two of us might well have concluded that all my father and I had in common were our balding heads as we bent them over the pages of the newspaper that morning in silence, each of us trying to make a bit more sense of how the world had changed and whether the future was turning out as well as we had been given to expect.

Had Lockheed prototyped its ideal engineer, the specs would not have fit my father. He was by his own admission a "crank turner" and not someone given to flashes of design brilliance. He was not a fervent hawk ready to argue that every defense dollar was well spent against Soviet treachery. He was not someone eager to do whatever Lockheed Missiles and Space Company asked, even when the deadline tied him to his desk on Christmas day, even when the work strained his arteries to the point of bursting.

That perfect Lockheed engineer was someone my father came to know in the mid-1970s, a stocky man with not much neck and a wide face, a genial if reserved fellow named Jerry. He and my father met on a black budget project in the Satellite Systems Division and they got along fine as far as their conversations went. They talked technicalese and they talked about the good bridge hands they'd played of late and that was about it. After months working side by side, my father was not even aware that Jerry had an oldest son, five years older than me, who was named Steve.

My father inferred merely that Jerry led an unremarkable Lockheed life. For the most part, he had. Like my father, Jerry had considered himself fortunate, when done with the Navy, to land a space-age job amidst orchard blossoms. Like my family, Jerry and his wife and three kids lived in a four-bedroom house on a street lined by similar houses. Like me, Jerry's son, Steve, belonged to a swim team and he grew up playing with other tanned and healthy children of aerospace.

But Steve was a boy unlike me. Nine years old, Steve tore through every Tom Swift book, reading them late into the night by the light of the street lamp spilling through his bedroom window. They were stories about a child engineer who builds robots and submarines and spaceships with his father. Already Steve had built a crystal radio with the kit his father had given him. Already Steve's father had given him other kits with buzzers and switches and relays and lightbulbs that could be hooked up thousands of ways.

In the fourth grade, Steve's father helped him create a science project showing which liquids conduct electricity. The next year, when Steve asked his father to explain the makeup of atoms, out came Jerry's old college textbook with the periodic table showing all the elements known to exist. Son and father decided to create a blinking shrine to these basic units of everything; they fashioned a display of ninety-two lights that switched on and off in combinations representing the electron orbits for each element. By the sixth grade, Steve had built a Heathkit ham radio and had earned an operator's license, as had his father. By then, Jerry was teaching Steve Boolean algebra, the zero-one mathematical language of electronics design, and Steve was diagramming logic circuits, having won a science fair competition with an addition and subtraction machine of his own invention. By then, Steve had transformed the street where he lived into a big connected circuit, stringing intercom wires along backyard fences and from light pole to light pole, wires that ended in his friends' bedrooms attached to microphones and lights and buzzers. Steve was nothing like me, the Lockheed son who was dreamy and not handy. Steve was Mr. Swift's son.

When dwindling contracts forced Lockheed to lay off legions of engineers as Vietnam wound down, the company looked for a way to keep one of its best. Jerry was given a rare grant to simply follow his instincts. He used computers to design and test integrated circuits so complex that Fairchild Semiconductor, an industry leader, showed interest in a joint venture. Lockheed balked, moving Jerry to a less experimental project, but the grant-funded work was unclassified, and so father told son about all the wonderful engineering he had been able to accomplish with the company's mainframe. That was the beauty of working for an institution so rich and established as Lockheed, mandated by the United States government to do top electrical engineering. Lockheed employees, if they were good enough, got to work with computers.

In those days, when Steve was a teenager, the son was fond of telling the father that he fully intended to possess a computer of his own, to have one in the next room wherever he might live. The father was fond of laughing and telling the son that a computer would cost him more than a house, so there would be nowhere to put it.

The son turned out to be correct, of course, because the son was that Steve who, at the age of twenty-four in 1976, invented the Apple computer.

When Jerry Wozniak's son did that, he did something more. He subverted the belief system of his father's tribe, a top-down, lock-step vision of how history-changing technology must come into being. Now there would be two technological peoples living side by side in the Valley, each vying to capture America's imagination with its particular idea of the future. There would be the blue sky tribe of aerospace, and there would be the tribe of Woz, a Lockheed son turned renegade.

The mythology of Woz and his miraculous machine has been burnished golden by countless tellings. We know how Woz met the other brilliant Steve in a garage when both were grade schoolers. And how the promotional chutzpa of Steve Jobs was drawn to the engineering genius of Woz. And how a drifting Woz was reinvigorated by kindred spirits at a gathering of computer hobbyists called the Homebrew Club. And how the Homebrew Club, founded by a peace activist who'd done time in jail, was chaired by a radical who put free access computer terminals in storefronts. And how the Homebrew firebrands yearned for a computer that might finally be used not against "the people" but by and for "the people." And how Woz answered the call with his Apples I and II, the latter creating a brand-new consumer market while catapulting Apple Computer into the Fortune 500 in five years, faster than any company in history.

The Woz of mythology is the man who won by daring to thumb his nose at authority (the authority revered by his father). Woz the boy planted shrill sound devices behind the classroom TV, made sirens blare in driver's ed. Woz put a metronome in a friend's locker and laughed to see the principal grab the ticking box and run outside with it close to his chest, sure he was saving the school from a bomb. "The teachers always said, 'We know Wozniak did this, but he's too bright to catch.' I never got caught,'' he once said. Woz, one quarter shy of an engineering degree, spurned the university and started Dial-A-Joke from his apartment. Woz, for a while, sold electronic ''blue boxes'' that illegally jimmied into phone networks; a pal was the notorious ''Captain Crunch'' who reveled in stealing long distance time from great corporations like AT&T.

All of this made me glad to know, nearly twenty years after the invention of the Apple computer, that finally I would be meeting the myth of Woz in person. This is why I returned to the Valley in the autumn of 1994. At the cluttered office he keeps in a nondescript business park just off the freeway in Los Gatos, Woz had granted me an audience. I was by then a Lockheed son who had spent much of my life seeking release from the customs and assumptions of the tribe of my father. What, I wanted to know, had made Woz do the same in such dramatic and legendary fashion? I was there, in a sense, to further polish the myth, to fit it neatly into my own private mythology of my tribe and its history. Why, I asked, did he so obviously relish challenging authority?

Woz seemed perplexed by my questions. He wanted me to know that he was a prankster by personality, yes, but not a rebel by ideology. ''Challenging authority. I don't like that phrase,'' Woz said. The wide face he'd inherited from his father, Woz continued to mask with a full beard, albeit a better trimmed beard than seen in all those famous photos from Apple's early years. ''I don't really like to challenge authority, but I don't like to accept something only because someone says they have authority. In other words, I really do like to be skeptical.''

He reminded me that before Apple happened, Woz the dropout, Woz the phone phreaker, had taken a job at Hewlett-Packard. This he'd done because designing circuits for calculators gave him what his father had: ''security'' in a big company with plenty of ''quality'' projects to hold his attention. ''I was so happy,'' Woz remembered, ''when other people at Hewlett-Packard said, 'You could work here for life and just be an engineer, not go into management.' ''

Populist rhetoric, Woz assured me, was not the draw that kept him coming back to Homebrew meetings; it was the room full of tekkies. The Homebrewers tipped Woz off to the availability of suddenly cheap microprocessor chips. Gratefully, he photocopied his evolving computer designs and passed them around for suggestions. And when Woz brought out an actual, working Apple I, the Homebrewers explained to him the miracle of his machine.

"All of a sudden the people in the club were talking along political lines, of individuals having these powerful tools that previously were kept from them, that they were going to be able to write better software than the big companies, that they were going to do more for the world and basically turn things upside down. . . ."

"Yes," I said (the legend coming back into focus).

". . . so there was a strong pro-individual type thinking in our club. And I loved hearing that because it was true!"

"Then it *was* true," I chimed in (reciting myth from memory), "that the Apple computer grew out of this very philosophy of yours. . . ."

Woz was perplexed again. "I have to think about that," he said, his tone that of someone who really does like to be skeptical. "No," he corrected me. "I just love computers."

Woz had never wanted to found a sociotechnological revolution, not even, at first, a company. That, he said, is why, even after the Apple I, he had continued to cling to Hewlett-Packard, to stay with his sure thing in a great corporation. Political talk certainly could not pry him loose. "In college I basically decided not to be involved in politics," said Woz. "The system goes on and we're going to have a really decent life no matter who's elected." Caring about politics, Woz had concluded, is a recipe for frustration, wasted emotion.

"I love computers. I love computing. I love any time I can get on a computer. . . ."

As I listened to Woz I was reminded of the conversation I'd had with his silver-haired father the day before. Jerry Wozniak had spoken in similar terms about his Lockheed work, absorbing work about which he'd never felt the slightest moral qualm, even when he had helped design the guidance system for a nuclear missile. He had exuded a startlingly pure attraction to the technical challenges Lockheed had offered him. He had confused me at the time by saying of his son, the mythical Woz, "He's done a lot of things I would never do, but I just feel kind of akin to him."

Jerry Wozniak had added that he felt not at all akin to Steve Jobs. The other Steve, whose parents were not blue sky but blue collar, was the one who finally managed to pry Woz away from Hewlett-Packard with pleading and cajoling. Steve Jobs did not find it presumptuous to imagine himself rich and famous and the boss of his own corporation. This flew in the face of Lockheed culture, a culture that expected modest egos of its employees, egos that fit neatly side by side in cubicle after cubicle, patient egos. And that is why Jerry Wozniak did not feel akin to Steve Jobs. That is why he remembered seeing in young Steve Jobs, in fact, a character flaw.

"He'd say, 'Start at the top.' Now, most people would start at the bottom and work up, you know? But he just never thought of it that way at all."

As Jerry told me this, we were sitting in the dining room of the house he and his wife had recently purchased after living for twenty-eight years in the house where Woz grew up. The new house was spacious but not ostentatious, a rather large ranch house next to other expensive ranch houses across the road from one of the last orchards in the Valley. "Steve Jobs just thought you should start at the top and we thought that was a character flaw." Jerry shrugged. "But he did it."

The father's instincts were shared by the son. Before founding Apple, Woz the loyal employee first offered his personal computer idea to unimpressed superiors at Hewlett-Packard.

The father's instincts lingered in the son long after he'd become famous. "Steve Jobs and I were very different. He just wanted to be at the top," Woz told me. "And I just wanted to be at the bottom. No one ever worked for me at Apple except one technician."

I heard in Woz the sound of myth crumbling. Perhaps the Lockheed son was not so different from his father. Perhaps the son had never, after all, consciously sought to turn his father's world upside down.

Was there ever a time, I asked Woz, when you could have seen yourself working at Lockheed?

"Oh, absolutely," he answered. "I just assumed Lockheed was electrical engineers like my dad, and from the time I was very young, I was going to be an electrical engineer. I knew I was good at it and I knew I liked it and I just figured, 'I'm going to be an engineer.'"

If sheer talent rescued Jerry Wozniak from Lockheed's early 1970s layoffs, my father believed that he and hundreds of other crank turners survived for a different reason. His life preserver was his security clearance. To bid on any new secret program, Lockheed needed to assure the government it had the personnel to do the work. Therefore management found it necessary, if not particularly efficient, to keep a lot of "ticketed" workers around even when projects were scarce. When contracts rolled in, the same imperative bent common sense the other way. Somewhere in the Lockheed complex there was a big room called "the ice-box" where perfectly able-bodied engineers spent weeks and months doing nothing but reading mystery novels and learning musical instruments and running their own side businesses over the phone. They were waiting for their clearances to come through. Until then, Lockheed paid them full salaries to kill time.

"Waste" and "efficiency" were relative notions in the culture of Lockheed, a culture that assumed government should guarantee profit even when jobs ran well over budget. "Cost-plus" contracting, it was called. When a project was done, Lockheed figured its costs, tacked on a profit "plus," and submitted its bill. Within this construct, the primary goal was to make products not cheaply but to military specifications, specs that called for back-up systems for back-up systems and for materials capable of withstanding a nuclear blast. Lockheed would build one or a few machines that marked the culmination of years of R&D, esoterica tricky, by its very nature, to price up front. Every time a part was respecified or redesigned along the way, that made other parts obsolete, threw the latest estimates out the window. Such a culture would seem to invite sloth and cheating and, of course, taxpayers have been angered by such revelations since the earliest days of aerospace. Jerry Wozniak, for one, never did understand what all the complaining was about.

"I've never seen this pork barreling stuff. This business about waste and how inefficient things can be." At Lockheed, he said, "I've just never seen it. Although it *is* hard doing business with the government. The tendency all the way up the line is: don't make a mistake, because you get penalized for mistakes. So we take extreme precautions to guarantee against making mistakes. And does that increase the cost? You bet!" But in the end, as Jerry saw it, the taxpayers always got what they paid for—even if they did have to take it on faith. "The stuff I worked on," Jerry assured me, "was *very unusual stuff.*"

Here, now, is where the mythology of Woz and his miraculous machine finds powerful allure, because *very unusual stuff* is precisely what Jerry's son managed to invent using a twenty-dollar microprocessor chip and twelve screws to hold the whole thing together. The first batch of fifty Apples was financed not by federal millions, but on thirty-day credit chits from parts suppliers. This gave Woz and Jobs exactly a month to hand-assemble their machines and sell them to a retailer and pay off their suppliers out of the proceeds. Myth and fact record that in Steve Jobs's garage the two friends soldered and wired like manic dervishes until, with a day to spare, they'd *done* it.

How much more thrilling is this story of creation than anything Jerry Wozniak could tell me, his work cloaked as it was in bureaucratized blackness? ("What did you and my father do together at Lockheed?" I asked Jerry. "We weren't in the plumbing business," was all he said.)

How much more thrilling to imagine *very unusual stuff* created by kids in a garage. More thrilling, as well, to see the marketplace transform, in a few short years, the very unusual into the quite usual. Unlike missiles or satellites or lunar modules, the personal computer could be touched, could be put to use by virtually *anyone*. Once we had, why shouldn't America prefer the myth of Woz to the myth his father (and my father) lived by?

Every now and then during the early go-go years of Apple Computer, Jerry Wozniak would drop by his son's company and shake his head in disbelief at what he saw there. Without days of meetings, without supervisors even checking off, "individuals were making decisions on the spot. They had to."

Jerry would compare this to what he had known all his days working in aerospace. "There's a limit to how independent you can be at Lockheed because if you're cutting against the grain too much, it is not going to work well for you, it is not going to work well for the others." Yet everyone at Apple seemed to run hard and even rudely against the grain when designs were up for grabs. And after decisions were made, well, Jerry had put in some long hours at Lockheed, but everyone at Apple seemed to be twenty-five and devoid of a home life, seemed to want nothing but to work around the clock to get the product out the door and onto shelves and into the hands of as many people as possible.

The Woz had taken what his father had taught him and he had made it his own and then he had given it over to a friend with the character flaw of a presumptuous ego. Now, between the world of the son and the world of the father, there was, Jerry concluded, "no comparison."

For some years after the invention of the personal computer, my family paid little notice to the tribe of Woz. When we finally woke to the commercial electronics boom in the Valley, we were confused by what came with it. The new tribe seemed to glorify impermanence, an idol we found hostile, even perverse. Their buildings were called "tilt-ups" because they were constructed of slabs of concrete, the slabs poured on the ground, then tilted up to make a place of work virtually overnight. Tilt-ups were populated by well-paid technicians who nevertheless scorned loyalty, "job hopping" every few years to other tilt-ups where more money or promotions awaited. Tilt-ups also housed instant assembly lines staffed by immigrant women easily fired when sales went slack or when a union drive (never successful) threatened. And tilt-ups were full of temp workers who came and went in rhythm to the short-lived "product cycles" of commercial electronics. By the early 1980s, Silicon Valley led the country in temporary employment.

My family found alien what the tribe of Woz caused around us, the rush to personalize the microchip and get it to the marketplace. Blue sky technology was built in secret for the state; its incomprehensibility was part of its appeal. But now suddenly everyone was a start-up schemer, racing to be the first to invent some new peripheral or interface, gadgets too "friendly" to too many "users" to hold any mystique for us. When the start-ups became big electronics companies, they still looked foreign to us. In their competition for the best minds, they held out odd visions of Utopia, "campuses" where ponytailed renegades ruled and everyone played volleyball at lunch. The same firms made a virtue of "lean and mean," farming out what they could to Third Worlders to whom they felt no obligation. To my family, this all seemed so un-Lockheed, so un-Catholic.

Interstate 280, with its signs declaring itself to be "The World's Most Beautiful Freeway," had been our freeway full of station wagons. But now 280 was crowded with Mercedeses and BMWs, the drivers spelling out their arrogance on personalized license plates, one WIZKID after another wanting us to know IMRICH. Now we read in the newspaper that families like ours had gone out of style: The Valley was a national leader in the rate of abortion, numbers of hours women worked outside the home, and the failure rate of marriage (well over half). We read of big deals sealed over cocaine and of assembly workers cranked on amphetamines (sometimes given to them by their supervisors). We learned that *work hard, play hard,* the code of the tribe of Woz, was driving people nuts. When one estimate suggested that two thirds of its workers were seeking some form of therapy, Apple Computer considered a money-saving move: Bring the psychotherapists into the workplace, under the company's own roof.

My family certainly did not see itself in the microchip millionaires, the start-up upstarts, the coke-bingeing burnouts, the way Silicon Valley was rewriting the career objective on its résumé. My family clung to the code of the military contractor, the Valley's pioneer code: sober reliability. Loosen up too much, and it might show up on the Lockheed polygraph. But pass every polygraph, and the badge you kept might insure you a job forever, might enable you to accumulate a comfortable pension as you lived snug in the folds of a corporation the United States government was determined to keep in business. Here is what we thought as we watched the tribe of Woz take over our Valley:

Why be ungrateful for a perfectly good enough life in the sun? Why risk wanting more?

When the front page said that the CEO of the Eagle computer company had died smashing his Italian sports car into a ravine on the day he'd taken his firm public, it seemed to us as fantastic—and cautionary—as an Old Testament story.

If Jerry Wozniak had been an accountant in Des Moines, Woz would not have invented a personal computer. Father and son agree on that.

"I probably wouldn't have got the little electronics kit for Christmas in the third grade, you know?" Woz said.

"An accountant wouldn't have talked about the things we

talked about," mused Jerry. "An accountant wouldn't even have known what Boolean algebra was."

In this lesson of father begetting son resides truth on another level, for the Pentagon commanded the computer into being. The Pentagon, as well, commanded into being the microchip that made a personal computer possible. This not a story the tribe of Woz likes to tell itself, for we are now in the realm of blue sky mythology, legends of government mobilizing great minds to create great machines in the name of national security, national survival even. These stories have been told by Dirk Hanson in his book *The New Alchemists*. They include:

The story of Vannevar Bush, dean of engineering at MIT and eventual head of federal defense research during World War II, who believed that "in a scientific war, the scientists should aid in making the plans." In 1930 he had seen the need for a machine that would calculate the path of an artillery shell. The computer of gears and cogs and rods that he and his team created they called a Differential Analyzer.

The story of British mathematician Alan Turing and ULTRA, the secret project he joined to decipher Nazi transmissions during the Second World War. Turing had theorized a machine that could think in Boolean logic. ULTRA's engineers and scientists, giving flesh to Turing's theories with vacuum tubes and photoelectric paper tape readers, cracked the Nazi codes and maybe won the war. They called their computer COLOSSUS.

The story of ENIAC, the computer begun in secret in 1943 by contract to the Army Ordnance Corps. Thirty tons, one hundred feet long, ENIAC used 18,000 vacuum tubes to create memory and solve equations in record time. The problems ENIAC crunched were those of shooting down the enemy and tracing rocket trajectories, problems of warplane flight simulation, problems having to do with another secret project of national security: the making of the atomic bomb.

The stories of ENIAC's postwar successor UNIVAC, whose development was funded by contracts from the Army, the Air

Force, and a company getting heavily into the intercontinental ballistic missile business, Northrop. The government bought the first three UNIVACs in 1948; not until 1954, when General Electric purchased one for data processing, was a UNIVAC owned by a private company. By then the bomb makers in Los Alamos had long owned their own version of the UNIVAC, which (fact and legend record) was named MANIAC.

The story of William Shockley and his transistor, technology that grew out of wartime radar research. The Pentagon was quick to see the benefits of solid state electronics as replacements for bulky, failure-prone vacuum tubes. In 1952 the Department of Defense bought nearly every one of 90,000 transistors manufactured by Western Electric and gave five million dollars for transistor research to General Electric, Raytheon, Sylvania, and RCA.

The story of the IC. The first integrated circuits on a chip, invented by Shockley disciples, were bought in the early 1960s almost entirely by the only customers who could afford them: the Pentagon and NASA and their contractors. Fairchild Semiconductor, which according to one of its executives achieved "liftoff" thanks to the Minuteman missile program, spun off many of the big chip makers in today's Valley, including Signetics, American Micro Devices and Intel.

The story of IBM. Since the 1930s, Thomas Watson's company had been supplying, as the name said, machines to business. But the company truly took off during World War II, growing four-fold, and at the dawn of the Cold War, Pentagon brass urged IBM to try to build a computer faster than the UNIVAC. The market for IBM's first mass-produced computer was the military's massive new radar network. That was the mid-1950s. By the end of the decade, IBM so dominated the government and corporate market for mainframe computers that an industry saying went: "IBM is not the competition. IBM is the environment."

To a Lockheed family, IBM, a maker of machines of the future, seemed to confirm our sense of how that future would be organized, groomed, and dressed. IBM was bureaucracy honed to ultimate profitability. From the command post in the green countryside, the perfectly landscaped headquarters in Armonk, New York, the orders came down the line. The uniformed army marched forward in polished black shoes and dark blue suits and white shirts and drab ties, a force with close-shorn hair and clean shaves, overwhelmingly Protestant and white and male. Within headquarters, where memos ran thick with secret code words and flip-chart presentations were locked up every night, security was an obsession. In the labs and in the branch offices, employees were given to understand they had jobs for life, even though that meant Armonk, in its paternal wisdom, might move them to any part of the country at any time.

IBM seemed to prove the strength inherent in monolith, the rewards for discipline and loyalty. In fact, IBM ran its own schools of acculturation; within those classrooms, as recently as the early 1970s, employees were taught to sing hymns descended from this "IBM School Song."

Working with the men in the Lab.,
Backing up the Men in the field,
Behind each one in the factory,
To a peer we'll never yield!

In every phase of IBM
Our record stands for all to see
The Alma Mater of the men
Who serve the world's best company . . .

I found these lyrics reproduced in a 1975 book called *And Tomorrow . . . the World? Inside IBM*. The author, a British business reporter named Rex Malik, sifted stacks of internal IBM documents and interviewed hundreds of IBMers at almost the exact moment that Woz was inventing the Apple computer in his garage. The conclusions Malik drew twenty years ago, he rendered with the funny, fierce sarcasm of a heretic thoroughly unimpressed with IBM's then spectacular bottom line and the many books then celebrating IBM's genius for modern organization. Foreshadowing what would become the accepted wisdom among today's corporate strategists, Malik disparaged the IBM middle manager, the quintessential Organization Man, as no prototype productive American. Enemy to his own creative potential and, ultimately, to efficient enterprise, he was little more than a well-kept soldier in servitude to an empire conquered by monopolistic methods. Malik damned corporate IBM as decidedly "militaristic, in that there are sharply delineated boundaries to the duties and responsibilities that an executive will carry, the whole is carefully ranked in rigid hierarchy, and any action undertaken likely to change the strategic picture one iota has first to be cleared upwards: it is not an atmosphere to encourage independent decision-making."

Malik furthered his case with this quote from a former Big Blue executive who had left to be president of a small company. "I know of no major decision I made in thirteen years at IBM that did not require six months of staff work. In my own business, no decision takes more than two days.

"I had a bizarre vision one night when I was thinking about my tombstone. It read 'When he joined IBM it was a struggling $500 million company. When he got his gold watch, it was a $40 billion corporation. We think he helped, but we're not sure.' "

This described the nightmare, of course, of a man who was presumptuous enough to want to be at the top, and so that man had left IBM for an uncertain life atop a small firm. IBM, at the time he did this, was a multinational juggernaut with well over 100,000 employees in America alone, 17,000 of whom were middle managers—17,000 gold watches to be given out someday, 17,000 eventual tombstones to be carved. What a believer in IBM might have said about that fellow's "bizarre" vision, what a

Lockheed engineer who was thankful to have survived the layoffs after Vietnam might have pointed out, is that a man could have far worse written on his tombstone.

"Totally, totally fun. New families had moved in and all the families had kids and we were the same age. There were kids all over, so many kids on our block, and we would just go up and down the block and run into each other and start riding bikes and agree to do something, go over to the school or something, and we had little clubs . . ."

Woz was remembering for me the feel of the neighborhood where he grew up. "Hardly anything existed and now every single thing has been taken for houses and shops and I miss that. That's why I kept moving further and further into the mountains. I've had three homes. But it's not the same. Because back then it was so great to be around kids. We were all great and sharp and bright and young. And lots of fathers worked in the same place, a company with the same sort of class orientation. It felt very *cohesive*, the neighborhood. The neighborhood was like a little island in the middle of orchards and I'll never see that again probably. Just like the start of Apple. It was an incredible time and I'll never see it again."

Since that incredible time, Woz had seen much. Ten years after the start of Apple, he had quit, saying, "I'm at a point where I want to work on anything but computers. I would be amazed if I ever make another computer in my life." He had married twice, both ending in messy divorces. He had learned to fly and had crashed his airplane, narrowly surviving. He had lost twenty million dollars on two rock extravaganzas, the US Festivals. He had traveled the world meeting leaders including Vaclav Havel and Mikhail Gorbachev. He had decided to become a schoolteacher.

By that time, the tribe of Woz needed only the myth of Woz, not the person. In 1980, six years after the start of Apple, a top strategist at IBM had concluded that Big Blue must get into the personal computer business, but that "we can't do this within the culture of IBM." Designers and marketers were plucked from the company's ranks and told to go off alone and try to emulate the culture of the tribe of Woz. They had come back with the IBM PC.

In the autumn of 1994, Woz was complaining to me that a father found it difficult anymore to buy his son an electronics kit. Heathkit had folded, and Woz was constantly looking for what Jerry had given him, kits with switches and relays and buzzers that could be hooked together a thousand ways. Woz said that bright children like his son, Jesse, who was twelve, spent their time in front of the computer screen manipulating software, not delving into the circuitry within the machine as Woz had. I asked him if he felt nostalgic for his youth, if he considered his own son in some way deprived.

"I just wish I had his life!" Woz answered. "I wish I were he because then I could grow up with computers and with video games and I know these things are so great that I would have just loved them, being the person I was. And I didn't have them." Already at twelve, Jesse knew more about programming the personal computer than did its inventor. The son, in fact, was teaching the father the latest tricks of software, how to design pages and troll the Net and download software updates and play new games. "Anything I can do on a computer, he does it faster. That's why," Woz told me, "I wish I were ten years old."

In this, Woz is different from his own father, because Jerry Wozniak never seemed to wish for his son's life. My father remembers that even after Woz became famous, Jerry said very little about him to colleagues at Lockheed, except to roll his eyes at the rock concert fiascos, saying his son had fallen in with the "wrong element" and paid the price for it.

Jerry's element was Lockheed and always would be. The grueling project in 1977 that forced him to work Christmas day did not dull his appetite, nor did the heart attack that ended that project for him. Even when the Cold War was over, even when layoffs were once again souring morale in the halls of Lockheed, Jerry Wozniak was eager to go into work. After he reached retirement age, he found a way back inside through consulting. And after his last consulting job was through, he quietly hung on to his badge. No one had thought to ask him for it. Until someone did, Jerry Wozniak would put on his badge and walk into Lockheed Missiles and Space Company. He would seek out his old colleagues and work on their projects for no pay. He would, at lunchtime, take his usual place in a bridge game that had been going on for fifteen years.

Jerry described for me his last day within Lockheed, a day not long before our conversation. He had spent the morning working on Missiles and Space projects for nothing, whiling away the lunch hour playing bridge with his friends. But a corporate vice president had caught sight of him in the hall, had done a double-take, and announced to all around, "What's *he* doing here?"

Someone took Jerry's badge from him. Then "they invited me to leave the complex." Jerry said he wasn't angry about it. "My guess is it was a liability problem or something like that. If I fall down the stairs, I guess it's not good if I'm not an employee."

It was several mornings after I spoke with Jerry Wozniak that I sat quietly reading the newspaper across from my father, waiting for the coffee to brew. I was thinking about how Jerry had defined to me his relationship with Woz. "Everyone in this family," he had said, "is fiercely independent." The son's skills and accomplishments had flowed from the father, but the father had not wanted to take credit. "I just pointed him to the right literature, answered a question here or there." The son had then, by luck and obsession and invention, constructed a life completely different from the father's, had made choices that made the father roll his eyes with disapproval. And yet Jerry had preferred to say of Woz, "I just feel kind of akin to him."

I was thinking that the son seemed to be triumphing over the father in the newspaper stories that morning, stories of aerospace decline in a Silicon Valley that was nevertheless prospering, stories praising aggressive young entrepreneurs and stories about the trimming of middle-aged, middle-manager workforces. I was thinking this as I turned to page 6b, which happened to be the obituaries.

The black headline at the top said: "Jerry Wozniak, Electrical Engineer."

Underneath, a smaller headline read: "*Father set example for Apple co-founder.*"

Jerry Wozniak had been at home in front of his personal computer when a heart attack killed him. The story said that Jerry was to have hosted a bridge game with "Lockheed pals" the next night. There was a quote from the famous son. "He was meticulous. I can remember him at home working and working, hours and hours on drawings trying to get a solution. I knew that feeling later when I spent hours and hours to save the tiniest bit of circuit." When his father was found dead, Woz reported, "Everything was neat, even his glasses were lined up straight."

I said to my father, "Jerry Wozniak died. Two days after I spoke with him. Can you believe it?"

"Jerry?" my father said, taking the page from me and reading the Lockheed life that had been compressed into eighteen column inches of some journalist's words, a blue sky life written about only because the son was the mythical Woz. My father shook his head as he read. I looked on, seeing across the table from me a face with rosy splotches and lines and stray white hairs, a face that seemed to have been made soft by history, including the history between the two of us.

Anyone who knew some but not enough about my father and me might well have concluded that our history was one of an example offered and rejected, opposites pulling apart, changing times opening a gulf between, so that now I was the son who (likely in some spirit of vendetta) made a living attacking the very basis of my father's own livelihood. But that is not really the way it had come to be with us. I knew that my father always had felt kind of akin to me and I to him and that this was something we understood in silent moments together like these. As I watched him reading the obituary of a fellow Organization Man that morning, I thought my father looked uncharacteristically fragile. And yet I could not imagine his face gone from my life, his face which would be my face, no doubt, in twenty-five years.

EIGHT

OUR PRESIDENT

Ronald Reagan made my family's tract house into a dream home. Just about the time of his landslide re-election in 1984, Ronald Reagan knocked out the north and west facing walls of our house. He tore out the linoleum floor and the flocked ceiling of the family room. He ripped out the frumpy cabinets and yellow tile and harvest-colored appliances in the kitchen. He moved the washer and dryer in from the garage, adding a laundry station next to the refurbished bathroom off the master bedroom. He cut a larger window in the living room. Then Ronald Reagan went to work on the new, pushed-out north and west walls and the additions they would enclose, the expanded family room with its high ceiling and free-standing fireplace and billiard table, the octagonal dining room that would be cozy yet plenty big enough for the family gatherings my parents expected to grow in size in the years ahead. Ronald Reagan gave my mother a kitchen twice the area of the old one, with solid oak cabinets and an extra sink and yards of counter space. He gave my father a shop of his own just off the garage. He got rid of every scrap of shag carpet in the house, sanding and lacquering the floorboards beneath until they gleamed golden. In the family room Ronald Reagan laid tongue-and-groove hardwood from Sweden, and in the kitchen and dining room, rich brown squares of Mexican terra-cotta. Everything shined with the light that poured in from many new windows, picture windows with wood casings instead of the original aluminum frames, a touch that elevated the final effect above the cost-conscious functionality the house once announced. Now the light poured in through picture windows that seemed almost ornate with their many panes (or, more precisely, the illusion of many panes, since the whole checkerboard of stained wood snapped out to reveal a single plate of glass, the easier to wash).

Ronald Reagan never actually laid eyes on my family's home, no. Most of the renovation was done by an independent contractor named Dick who probably voted for Ronald Reagan and who probably bid too low on my parents' project. Whenever my father was niggling with him over some detail in the work, Dick liked to flash his smile, white against tanned leather, and say, "Hal, building is an art, not a science." So it was Dick and his small crew who did most of the work, and it was my father who added many finishing touches of charm, including the Swedish hardwood and the Mexican tile and an inlaid redwood ceiling over the dining room.

But it was Ronald Reagan who wanted my family to have all this, Ronald Reagan who arranged to have it paid for. In fact, no president of the United States ever did more for my family. And so it may seem strange if I say that my family accepted Ronald Reagan's every blandishment, yes, but did so the way a child takes a Christmas gift from a detested old uncle, with eyes averted and confused feelings—feelings of covetousness and resentment and entitlement—swirling inside. To my father and mother, liberal-minded people, Ronald Reagan was a sham whose politics were mean-spirited. To me, a young man who wished to imagine his soul cleaner than most, scrubbed by indignation and good works, Ronald Reagan was something more. He was the devil himself, denying me my self-satisfaction, reminding me always of my complicity in his schemes, the complicity of a child so fortunate as to have been born into the blue sky tribe, the tribe Ronald Reagan loved and favored like no other.

Surely no president ever professed more belief in the powers of aerospace invention, a vow of faith that Ronald Reagan, over and over again, invited all of America to renew. He did this by returning the mythology of holy crusade to the Cold War, a crusade that was to be won, inevitably, in the skies and in space. Years of "detente" as official policy had drained the drama out of the Cold War story, had made the Cold War, under President Carter, a matter of keeping our Olympic athletes home and off TV if the enemy should happen to invade Afghanistan. It took Ronald Reagan to undo what President Ford had done when he called whipping inflation the moral equivalent of war, to undo what President Carter had done by trying to make driving fifty-five the moral equivalent of war. It took a President Reagan to restore the *Cold War* to the moral equivalent of war, to brand the Soviet Union an Evil Empire, to joke darkly about bombing Moscow, to prophesy that Armageddon was nigh, that a Christian nation must therefore prepare itself to engage and vanquish the pagan enemy.

Ronald Reagan came into office the first president to speak openly of waging and winning a protracted nuclear war. "Winning," by some White House estimates, meant obliterating the Soviet Union while suffering twenty million civilian casualties at home. In making this his test of victory, Ronald Reagan broke with the doctrine of Mutual Assured Destruction (MAD), the assumption that neither superpower would start, much less win, a nuclear war, because once the thousands of missiles on each side started flying, holocaust would devour both sides. There were several ways to read the Reagan shift. Perhaps the man *did* so hate and distrust Soviet Communism that he spoiled for a showdown no matter the cost. Or perhaps Ronald Reagan was more canny than that, meaning only to prevent the Soviets from assuming they could render us helpless with a first strike, in which case he merely aimed to shore up the logic of MAD. Or, more canny still, the missile rattling could have been bluff intended to drive the Soviets into economic ruin as they strove to keep pace. No matter. Whatever intent lay behind all this Cold War mongering, my family was sure to benefit, for as Ronald Reagan set about mounting the largest peacetime military buildup in American history, he was commanding into being some wondrous, and wondrously expensive, inventions of aerospace. There would have to be computers faster than any before, capable of selecting targets well after their human masters had been incinerated. There would have to be communications satellites unlike any before, capable of withstanding the barrage of circuit-frying radiation that nuclear explosions send thousands of miles through space. There would have to be B-1 and Stealth bombers and Trident and Pershing and MX and cruise missiles, and there would have to be 16,000 new nuclear warheads added to all these various new and improved "delivery systems."

The problem with Ronald Reagan's reinvigorated version of the Cold War story, of course, was the implied dark ending for the victors, twenty million dead or maimed and the rest of us cowering as blue sky weaponry performed wondrously above. The aftermath "would be a terrible mess but it wouldn't be unmanageable," Ronald Reagan's director of civil defense, trying to put the best face on it, assured the writer Robert Scheer. But nothing is harder to summon than faith without hope. And so . . .

"Let me share with you a vision of the future which offers hope."

On the evening of March 23, 1983, Ronald Reagan once again invited a renewal of faith in the powers of aerospace. This time he defined hope in terms of meeting the ultimate systems engineering challenge. A shield in space, he asked us to imagine, impervious to nuclear attack, hammered from the stuff of blue sky dreams. "Let us turn to the very strengths in technology that spawned our great industrial base and that have given us the quality of life we enjoy today."

The famed "Star Wars" speech never mentioned Star Wars or even the official term for Ronald Reagan's vision of hope, the Strategic Defense Initiative. Ronald Reagan simply let it be known that he had been mired in nuclear gloom of late and that "my advisors, particularly the Joint Chiefs of Staff, have underscored the bleakness of the future before us." Fortunately, those same advisors and Joint Chiefs of Staff were in agreement with Ronald Reagan that the future need not be bleak at all. Not if "the scientific community who gave us nuclear weapons" could be commanded to work on "the means of rendering these nuclear weapons impotent and obsolete."

True, this space shield would take "probably decades" to construct. "But what if free people could live secure in the knowledge that . . . we could intercept and destroy strategic ballistic missiles before they reached our own soil or that of our allies'?" asked Ronald Reagan.

And true, this space shield would be costly. "But is it not worth every investment necessary to free the world from the threat of nuclear war? We know it is!" declared Ronald Reagan. He wrapped up by straining for chords once sung by John F. Kennedy, echoes of a time, nearly forgotten, when aerospace people had first been invited to imagine themselves arbiters of America's future. "My fellow Americans, tonight we are launching an effort which holds the promise of changing the course of human history. There will be risks, and results take time. But with your support, I believe we can do it."

In the weeks and months following his Star Wars speech, pictures began to emerge of what Ronald Reagan's space shield might be. Seconds into an attack, our supercomputers would launch hundreds or thousands of satellite battle stations which, once in space, would target the flaming boosters of rising enemy missiles and knock them from the sky with killer beams. The pictures came to mind sharp edged and chiaroscuro, like Chesley

Bonestell paintings and George Lucas movies. The words were technicalese raised to incantation, the prayers of crusaders whose eyes were lifted toward the heavens. The saving killer beams, America was told, might be streams of neutral particles. Or they might be free electron lasers shined from Earth to bounce off mirrors in space. Or they could be X-ray lasers "pumped" by a nuclear explosion aboard each satellite, this last idea given the name of Excalibur.

"You wouldn't believe. The money just *gushed* in." That is how my father remembered Lockheed Missiles and Space Company right after Ronald Reagan assumed the presidency. "I can think of more than a handful of programs that sprung up literally overnight and ended up costing billions of dollars. Ideas that had lain dormant for years, ideas based on unproven technology, were suddenly given the green light on a handshake and a promise. We couldn't hire people fast enough."

After my father was assigned to one such program, he spent more and more time away from home. Often he stayed in Lompoc, a little town next door to Vandenberg Air Force Base, a launching pad set in the cattle country just up the coast from Santa Barbara. Other times he was gone for days and weeks to a place that mysterious people on the phone called The Ranch.

"Hal there?" an extremely serious male voice would ask whoever picked up the receiver at my parents' house.

"No. Can I tell him who called?"

"Tell him Gunner called. From The Ranch. He'll know." Click.

What was this Ranch where Ronald Reagan had created new work for my father and for "Gunner" and for how many more? My mother and her children were curious, of course, but we had only the slimmest of details with which to construct a mental picture. We knew a man could find himself in some very high and precarious places at The Ranch, because one time my father returned wearing a strange pair of glasses, clunky plastic frames bought off a drugstore rack. He had lost his, he said, "while stepping onto a catwalk. I bumped my head and off came my glasses. I heard them hit the floor about, oh, eight to ten seconds later." My father smiled as he said this, smiling as he tended to smile when he had just told you something that was very intriguing but just shy of violating his security oath.

We knew The Ranch was in a place that could be very dark, because another time my father came back with a scabbed cut in his forehead. All he would tell us is that he had been driving across some dim landscape in the middle of the night in a rental car with the lights off and he had run into something and his head had been thrown forward into the steering wheel. "Why were you driving in the dark with no lights?" his wife and children all wanted to know. But his answer was a smile.

We knew that whatever went on at The Ranch made my father less prone to the sullen spells on Sundays before another week at Lockheed. In fact, he often seemed excited again to be an aerospace engineer. For twenty years he had sat behind a desk writing up test reports until he had all but forgotten what had drawn him to Lockheed Missiles and Space Company in the first place, the intimacy with exotic and even heroic inventions he had expected. After all that time, my father had come to see himself as one barnacle among thousands, firmly stuck to the belly of a great company whose guiding brains were not even aware of his existence.

The gushing is what washed him loose. All those Ronald Reagan dollars flowing into Lockheed created the need for new project leaders, and my father soon found himself head troubleshooter for whatever black budget technology was being refined at The Ranch. Now he was responsible for bringing together that system's designers, builders, and testers and directing everyone through the final debugging process. To my father's relief, he finally was spending much of his day with actual machinery that needed his corrective hands upon it. Often, too, the job placed him before audiences of Lockheed and Pentagon brass who listened to his flip-chart presentations about how life at The Ranch was progressing. Apparently my father shined in both roles, for he was promoted into management, and many times during the Reagan years my mother let me know that my father had received yet another raise or performance bonus.

In his own wry words, my father came to be seen "as something of a testing guru" within the Satellite Systems Division of Lockheed Missiles and Space Company. "Eventually," he said, "I became the trusted spokesman for such matters. It just turned out that I seldom was challenged because I turned out to be right so often." (This is the closest I have come to hearing my father boast.)

Versions of my father's story, repeated thousands of times around tract home dinner tables all over America, made aerospace communities vote overwhelmingly for Ronald Reagan. And why wouldn't they? How whole and complete, how self-affirming must it have felt for those aerospace workers to put an X next to the name of the candidate who so ardently endorsed their chosen way of life. How different, though, was the feeling for my father and mother, who happened to be among those many Americans who could not make themselves believe much, if anything, that Ronald Reagan ever said. My father had long disparaged the "hucksters" of Hollywood, and was not Reagan one of those, applying his same oily methods to politics? My mother had found some measure of grace in any political figure who was "for the poor," but who could trust a man who blamed the poor for their own misery? During his eight years as Governor of California, my family had come to see Ronald Reagan as somebody else's governor, but certainly not *ours*. Now Ronald Reagan had convinced enough other people to make him somebody else's president, but certainly he was not *ours*.

We cringed when the new president said on television, with straight face and twinkly eyes, that America could spend a trillion more dollars on defense while lowering taxes and cutting the deficit, all of this thanks to the Laffer Curve, a revolutionary new product that made economies grow faster than ever in history. We did not believe in the Laffer Curve, and we marveled that anyone else could. We did not believe that trees gave off pollution or that most people on welfare were lazy cheats or that most homeless people slept on the streets because they wanted to. We did not believe America could survive and win a nuclear war, nor did we believe that a space shield could be made to preserve us. We believed that Ronald Reagan was either a con artist or the slow-witted tool of con artists around him. In either case, his generous gifts to us would have to be morally tainted, ill-gotten loot thrust into our hands.

But what was one to do when the old uncle you can't love has placed his hands around yours and, with eyes twinkling, is closing your hands around what he wants you to have, a gift stolen from someone else? For that, in the end, was the triumph of Ronald Reagan: not the slashing of taxes or government, neither of which shrunk significantly under his direction, but the massive transfer of federal subsidy from one group less fortunate to another very fortunate indeed. What he took from the smokestack cities he gave to the blue sky suburbs, building dream homes for families like mine wherever military contracting was done, along Route 128 outside of Boston, in Grumman's Long Island, in Florida and Texas and Boeing's Seattle, in the flourishing beltway around Washington, D.C., where all the Pentagon lobbyists and consultants came to roost, and most particularly in Ronald Reagan's own state of California where nearly a third of all defense spending flowed, most of that to Orange County and Los Angeles and Silicon Valley. One in eight workers in California held a defense-related job and for these favored ones, for families like mine, The Great Communicator proved a very great appropriator.

When John F. Kennedy had smiled so brightly upon our tribe, he had spun an illusion that presented my family, a Catholic, liberal-minded aerospace family, with nothing like this moral dilemma. He had gushed military spending into aerospace, true. But at the same time he had charged us with a believably benign crusade, a race to the moon, a national excursion to space that would leave no one behind, not the poor, who would benefit from new Democratic social programs, not the blacks, who would live in a desegregated America of opportunity, not even the peasantry of backward nations, who would be lent, out of the goodness of our hearts, our Peace Corps know-how. The newly arrived aerospace middle class need not imagine itself distinctly favored; the country seemed rich enough, generous and expansive enough, that families like mine could tell ourselves we were only reaching the sunny suburbs a little ahead of the poorer folk who someday would join us there.

Ronald Reagan, for all his spinning of illusion, for all he gave us, offered no such comfort to the conscience. Ronald Reagan's words and policies rarely made Kennedy's pretense of binding the nation. Ronald Reagan's ideology tended to divide America into the deserving whose fate need not be tied to the undeserving, the winners who owed nothing to the losers. As the anointed winners, not only were aerospace families handed the spoils from dismantled Great Society programs begun by Kennedy and Lyndon Johnson, but when that could not cover the Pentagon outlay, we were paid with funds wildly borrowed, money lifted from future generations. As Laffer Curve collapsed and recession took hold and federal deficit ballooned, Ronald Reagan continued to finagle for us with the shamelessness of a Buy On Credit! telemarketing pitchman. Four years into his reign, defense spending was up 60 percent, to more than $250 billion annually; within a few more years, the levels of the mid-1970s had doubled, surging past $300 billion. Even better for the blue sky tribe, each new Pentagon budget reserved a growing share for research and hardware, aerospace specialties. By the time he was through, Ronald Reagan would have America spending a third more on hardware and R&D than even during the peak years of Vietnam. All the while Ronald Reagan, eyes twinkling, would laud the inherent "magic of the marketplace," as if he had not artificially rejiggered the economy to enrich my tribe; as if, as *Fortune* noted, the same federal investment in other sectors would not have created 25 percent more jobs, albeit for different Americans; as if *Businessweek* did not acknowledge the reality with headlines like this one, printed a year after Ronald Reagan's landslide election to a second term: "Pentagon Spending Is the Economy's Biggest Gun."

That, in the final analysis, is why my family so resented Ronald Reagan even as he gave us our father's revitalized career, our dream house, our secured place among the winners of the 1980s. We resented him for not granting us the one illusion we wanted most: a way to continue believing that our gain did not come at the expense of other Americans, their suffering, their diminished futures. Instead, we were forced to concede that Ronald Reagan, beneficent con artist, spokesman for Armageddon, was very much *our* president whether or not we gave him our votes, and the question became how to make personal peace with that fact.

In the spring of 1984 I found myself at the wheel of my parents' station wagon, my brother, Dan, next to me, my father in the back seat with my mother and my sister Maggie. We were heading north on Highway 101, having spent the weekend in Santa Barbara, Ronald Reagan's chosen home. We had been there to enjoy the casually affluent charm of the place and to see my sister Marybeth graduate with a bachelor's degree in ergonomics from the University of California at Santa Barbara. We had attended Mass in the old mission and we had window shopped along ritzy State Street and after the graduation ceremony we had looked on with delight as Marybeth and her friends danced on the beach in celebration, spinning and falling into the sand as the B-52s, singing "Rock Lobster," blasted from an apartment window.

Now, as we sped north on 101, there flashed into view a

billboard urging travelers to turn off and visit the town of Lompoc, a beautiful spot, said the sign, ". . . where marigolds and missiles mix." All of us laughed at that, knowing as we did that my father was spending a lot of time around Lompoc, and not for the marigolds. "Lovely Lompoc! Where nasturtiums nuzzle the nukes!" I said, eyeing the rearview mirror to catch my father's chuckle, seeing his smile fade away when someone, my sister, I think, asked him whether he did in fact work on nuclear weapons.

"Directly? No. But I cash every paycheck Lockheed writes me. And Lockheed is a company that is in the business of preparing for nuclear war." But for the humming of the tires on asphalt, the car grew silent and I began to feel bad that I had invited the chill with my joke.

I said, "I think it's a pretty good guess to say you work on satellites, Dad. Satellites that take pictures that make it possible to verify nuclear treaties." In the rearview mirror my father's eyebrows were raised with amused indulgence. "So, really, Dad, what you do *prevents* nuclear war . . ."

"Let's say I did happen to work on such projects," came my father's voice, calmly arriving over my shoulder. "And I'm not saying I do. Have you ever considered that the same satellite used to verify a treaty might also be used to pinpoint enemy targets for an all-out first strike? With the aid of satellites, those targets were picked long ago and they are constantly being updated. Insane, I know. But you can bet on it. And now there's a man in the White House who seems rather, shall we say, *cavalier* about pushing the button," said my father. "So if you ask me, these days nobody at Lockheed is clean."

I was out of my depth, I saw. My father had obviously thought about such things a lot more than the rest of us, and he, for one, had decided that it was better to name his complicity than to deny it. Everyone else in the car just seemed to want the conversation over with.

"I'll tell you one thing, though," my father said as we rolled on toward Silicon Valley, home. "I haven't noticed our bills getting any smaller lately. I'm going to keep cashing those checks."

I suppose I wanted to believe that the father's complicity need not be visited upon the son, that my aerospace upbringing was largely immaterial to who I was and who I could be and that I was free, therefore, to make a life cleanly separate from Ronald Reagan's designs. I suppose, too, at some level, I believed in the expiation of Original Sin through good works. Frankly, I am not sure how I explained it to myself back then, when I was twenty-two and deciding I should become a missionary of sorts to black children in a ghetto. When Ronald Reagan was first elected I was one year out of college, unemployed, and facing a move back into the family tract house. The world seemed not to care a whit that I had a bachelor's degree in English from small, Jesuit-run Santa Clara University. The world knew nothing of the self-regard I'd accumulated as a student, an insouciant, even lazy, student much of the time. That self-regard, I knew deep inside, was in too small reserve to withstand the move home (the oldest son back in his bedroom at the end of the hall at an age when his father was flying fighter jets off the deck of an aircraft carrier).

I found my escape into my mother's institution, into the Catholic Church. I signed up for a year of work with the Jesuit Volunteer Corps, an organization modeled on the ideals of Dorothy Day's Catholic Workers: activism for the oppressed joined to an inner life of serious prayer. The inner life of serious prayer held no interest for me, since I no longer practiced Catholicism and preferred to call myself, having read a few books by Existentialists, one of those. But the work with the oppressed appealed to me strongly. The oppressed, at the time, were Ronald Reagan's oppressed, and I found something exhilarating about the uncomplicated badness of Reagan, his toadying to crass wealth, his affable bullying of the weak, his easily caught lies. Rarely does a president of the United States afford a young man such opportunity to burn the flame of outrage so hot and clean. Besides, I felt teaching black children in the ghetto might add texture to my suburban character. And so I went to live on the edge of Hunters Point in San Francisco with six other Jesuit Volunteers in a drafty old pile of a house that nevertheless provided me a home warming to my self-regard.

We were three ghetto teachers, two prisoner rights lawyers, a campus organizer, and an office worker for the pro-Sandinista Institute for Food and Development Policy. We were all white and middle class and in our twenties, all of us given shelter, simple meals, and fifty dollars a month by the Catholic Church in exchange for our good works. Other Jesuit Volunteers across the country were doing similar good works in those awkward months between college and career, teaching on Indian reservations and tutoring Haitian immigrants and running workshops for the mentally handicapped. We were unquestionably for the poor, and so whenever we assembled for one of our religious retreats or blowout parties, it was understood that all of us were against Reagan.

Unfortunately, as a ghetto teacher I proved to be no Sidney Poitier. The kids ran roughshod over my friendly teaching style, called me "Mister Pointynose," accurately sensing that I was not in this thing for the long haul. When I took some of my students roller skating, I led them down a too steep path and one boy broke his arm. When I took some of them to a Giants baseball game, I slathered myself with sunscreen and never bothered to think that black children sunburned, too. When they woke up the next morning whimpering in pain and their mothers asked why, the children answered that Mr. Beers did not share his lotion. The principal fielded irate calls all that day.

Still, whenever I'd ride the commuter train south, whenever I'd arrive for a Sunday dinner, my mother and father and siblings were eager to hear my stories of simple living and ghetto teaching. In honor of the occasion my father often would barbecue a "tri-tip," a steak recipe he'd discovered in Lompoc on his Lockheed business trips. The salt-encrusted beef would come off the grill oozing black-red juices and there would be Napa Valley cabernet and many delicious dishes made by my mother who had trained herself to be a nutrition-conscious gourmet cook. Inevitably talk would come around to Ronald Reagan and our heads would shake at his latest outrage. My father would apply his readings about general semantics to the "pap" spoken by the president's officials. My girlfriend, Deirdre, and I, enamored with leftist insurgencies in Central America, would chip in our well-footnoted critiques. My mother would nod her amens. She was then a close reader of the magazine of the Maryknolls, missionaries whose involvement with movements of the poor often made them targets of death squads. The magazine on the living room coffee table was full of brown faces, the same brown faces that had once stared piteously from Catholic Charities envelopes at Sunday Mass. Now, though, the brown faces were said by *Maryknoll* magazine to be not pitiful but brave and determined, struggling to create a farmer's cooperative or a union drive in defiance of one of Ronald Reagan's dictator friends. The Maryknollers had helped my mother become comfortable with words like "liberation" and even "revolution" whenever other Catholic words, words of religious faith, were joined to them. And so, at dinner, as the second helping of tri-tip was served, it was possible for me to imagine that my father, mother, Deirdre, and I were all speaking the same language.

There were times, as a new bottle of cabernet was retrieved and uncorked and poured, when I heard all of us speaking a language as radical as anything said at the Jesuit Volunteer house in Hunters Point. Indeed, one evening my mother and father decided to pack their big kettle of a grill into the back of their car and drive north, arriving at Hunters Point with two tri-tips in the cooler. That night my father barbecued his steaks for my radical Catholic friends, and even the vegetarians took some bites as we all shared laughs at the expense of Ronald Reagan.

After a year, having flunked at ghetto teaching, I left the Jesuit Volunteer Corps to be clean of Ronald Reagan in what I hoped would be a more a dashing role: as a muckraking journalist.

In need of a first break, I made my way to the jungle regions of Chiapas, Mexico, where tens of thousands of refugees from Guatemala were gathering. Most of them were Maya Indians who had seen their villages destroyed by government soldiers carrying out, with the blessings of Ronald Reagan, a "scorched earth" policy against supposed guerrilla sympathizers. The escapees were stumbling into Mexico's remote borderlands, creating a crisis for the Mexican government. I learned of one desperate encampment just across the Usumacinta River from Guatemala, a mere dot in the rain forest reachable only by plane and closed to all reporters by fiat of the Mexican government. Conditions were terrible there, the Catholic aid workers told me, and someone should expose this. I bribed a pilot to fly me into the camp along with another man who would interpret for me. After three hours of picture snapping and interviewing, we stood on the runway waiting for the pilot to return for us as he had promised he would. He never showed. As a result I had much more time, days, to speak with the Maya, to listen to them tell me in their singsong voices of the tragedy they had experienced. I wrote their stories in my spiralbound notebooks and photographed their expressions of contained grief as they told of seeing husbands and fathers rounded into churches and shot, sons and daughters hung by their wrists and bayonetted, babies burned alive in lime kilns. I ran my viewfinder over their hands and eyes and lips, waiting for their pain to best announce itself to film. I jotted stars next to the most compelling phrases.

They hunted us like animals.
We are like children without a place to lay our heads.
We are lost little birds praying that God will hear our cries.

In seeking this pleading poetry for my notebook, I was pulling them away from their labors of survival, the lashing together of hundreds of huts from jungle thatch, the setting up of schools in four Indian languages, the organizing of a democratic system for sharing what little food there was. They were giving me their waning energy because I had come on an airplane with recording devices and a white face that other white faces might pay attention to, and this filled me with an incongruous joy, a sense of perfection of purpose. After acquiring more than enough stories for my needs, however, I began to feel vaguely ashamed to ask for more, to want to just leave the camp as the Maya knew I soon would (and they could not). The chance came after three days when another Cessna landed to collect mahogany logs. I paid the pilot fifty dollars to fly us out with the precious lumber.

At home, as the brown faces came up in the photographic developing solution in my darkroom, I heard, for a short while, their singsong voices pleading again. Then a different voice told me I should hurry for fear that someone might scoop me, and the last twenty pictures were necessarily a whirl of production. The May 8, 1983, edition of the *San Francisco Examiner* carried my report on the front page, jumping to three more pages inside with thirteen photographs accompanying. This was an excellent "clip" (as freelancers call their product) and it proved the break I had hoped it would be, landing me further assignments on a variety of topics.

Freelance muckraking did not begin to pay my rent, however, so I spent much of my time on assignments of a different kind. A silver-haired public relations man named John hired me on contract to compose brochure copy and advertisements for his client Transamerica Delaval, a manufacturer in Oakland, California, that sold diesel generators to the military and the nuclear power industry. I enjoyed visiting Transamerica Delaval's foundry, the largest west of the Mississippi. I would put on a hard hat and watch the men, many of them black and layered over with blacker soot, as they fed blast furnaces and pulled levers that sent cascades of molten, glowing metal pouring down from mammoth buckets. It was like visiting a working museum of fading "industry," and when the air grew too thick and hot in the dark cage of a factory, I would stroll over to the offices of the suit and tie men in order to discuss my work, the latest promotional copy I was writing for management. They liked me to write in a tone of gruff confidence, even though Transamerica Delaval was going broke, the foundry laying off more and more workers, everyone hanging on to the hope that Ronald Reagan might yet save their futures by coming through with the five-hundred ship Navy he had vowed as part of his mighty arms buildup.

After an Oakland visit I would return to John's office in pretty Portola Valley, a place of golden hills and country estates, a place where venture capitalists tended to live. John, a Republican who considered *Businessweek* the ideal filter on events, was kind enough to give me, free of rent, a corner of his office where, for forty dollars an hour, I sang the glories of diesel generators and, also for John, the glories of printing factories and of oil pipeline compressor valves and of a hand-held computer that let you calculate your tennis shot percentages. Often, over lunch, John and I would argue about Ronald Reagan. Then, if the work was light and the day was warm, we would take to the tennis court behind the office, our games almost always close. We became good friends. On weekends John sometimes invited Deirdre and me for sails on the boat he kept in Sausalito. We'd cross the Bay and tie up at the dock of Greens, a Zen Buddhist-run restaurant with polished redwood burls for tables and grand windows framing the marina and the Golden Gate Bridge. The vegetarian breakfasts, expensively nouvelle, attracted a clientele who were clearly doing well for themselves in the early 1980s, people free to make whatever they wished of a bright, crisp weekend day in San Francisco, people who were beginning to be called "yuppies." Sitting in Greens with John's boat bobbing in view, I thought of myself as an indifferent dabbler in their world, my soft sweater tied around my neck like a disguise, my self-awareness separating me cleanly enough from the yuppies who must have thought I was one of them.

A year after my visit to Chiapas, I returned to write a second (and as it turned out, final) newspaper piece about the plight of Maya refugees. Then Deirdre and I traveled on to the Caribbean island of Barbados, where she had a grant to study sweatshop conditions in the region. As if to help along my journalism career, Ronald Reagan ordered an invasion of the next island over, Grenada, but my reporting of the aftermath was uninspired and I found few homes for my pieces. Living with Deirdre on Barbados and then the nearby island of St. Lucia, I seemed to invite puzzlement and mistrust from my black and East Indian neighbors, the same cautious curiosity shown me by the ghetto children I had once asked to call me "teacher." I wanted to write movingly about their folkways and tropical wisdom, but willing myself into their lives was not going to be enough, clearly, to allow a communing of spirit. Not if I could not eat blood pudding and pig organs with them on New Year's Day. Not if I could not cheer on a cockfight or ease through an afternoon of joking and musing in the rum shack at the end of the road. Not if I could not make their children laugh and want to be in my company. My stiff attempts at all these things were usually seen for the calculated efforts they were. Perhaps I was CIA, rumor had it in the little village where Deirdre and I rented the best house and my electric typewriter filled the air with gunshot *rat-a-tatting*. Perhaps I was a cult missionary or some odd version of a rich tourist, since I did not seem to work for a living. In truth, I was hopelessly the American child of aerospace, just beginning to glean the limits to suburban-bred hubris. But how to explain that to the man across the way who slaughtered his favorite fighting cock to impress me and then, in vain, searched my face for authentic pleasure as I chewed on the stringy drumstick he had ordered his wife to cook?

I left the Caribbean knowing, without telling myself or Deirdre, that I lacked the appetite for life with Ronald Reagan's oppressed in the Third World. Back in Silicon Valley, my middle-class imagination now seemed only to want to write about people in the middle: bright, educated people struggling to live cleanly amidst Ronald Reagan's seductive offerings. I thought I might have found a model of success, one February morning in 1986, in the fascinating person of Pierre Blais.

Pierre was a "knowledge engineer," an expert in artificial intelligence at a moment when most AI research was paid for by Ronald Reagan's Pentagon: $600 million were earmarked for a crash program to create "intelligent" weapons right out of a science fiction. The Army was to get a tank that drove itself. The Air Force was slated for an electronic "pilot's associate" that conversed in English as it selected targets of destruction. Navy and Army were each to have their own "battle management system" keeping tabs on combat and advising strategy to commanders. Should nuclear war break out, artificially intelligent computers would say when and how to launch our missiles and the Star Wars defense. For the morning after a bout of NBC warfare (Nuclear/Biological/Chemical), the Army wanted a robot smart enough to roll through a body-strewn battle site and "detect and identify NBC contaminants, decontaminate human remains, inter remains, and refill and mark graves." Another Army report looked forward to the day when corpse-gathering robots "could be loaded and, by merely activating a switch, dispatched to the nearest mortuary."

The thematic richness of this macabre world was lost on neither me nor Pierre Blais, who could quote verbatim from Mary Shelley's *Frankenstein* and who likened his profession to "the unquestioning scientists under Hitler. Whatever destructive technology you create is eventually going to be used—and that goes for artificial intelligence." Pierre, sandy-haired and fortyish, a wearer of wire-rimmed glasses and double-knit suits, spoke of his troubled conscience in the precise tone of a technician, a tone cut sharper by a tinge of acid, a tone not unlike my father's whenever he chose to ruminate aloud about compromises made, complicity accepted. Unlike my father, however, Pierre claimed to be finally, truly, a happy man, having shaken loose the devil of Ronald Reagan once and for all, and I wanted very much to write a story about such a victory.

The more we talked in that first interview, the more it became clear that Pierre Blais had suffered a lifetime of sensationally misplaced faiths. As a teenager he had left his native Canada to fight on the side of righteousness in Vietnam, joining the Army's 101st Airborne and winning a Bronze Star with a "V" for valor during the Tet offensive. But by the end of his stint, he would weep while standing guard, his head full of horrors he'd witnessed, piles of dead civilians "stacked like cord wood" on the road to Hue, a captured enemy executed before his eyes while Pierre, paralyzed, stared at a confiscated picture of the man's family. Pierre had come home from that failed war with a "philosophy of pure nihilism plus pure pessimism, but I wasn't ready for that. I was too young."

And so he had joined the Mormons, married a fellow convert from Okinawa, and enrolled at Brigham Young University in Provo, Utah. There he learned computer programming under CIA consultant Richard Beal, soon to be Ronald Reagan's special assistant for national security. By 1980 Blais himself was helping to program the CIA's computer nerve center, working under a top secret clearance for Logicon of Southern California. The work was stimulating and he had given his wife and three daughters a cozy life in the suburb next door to CIA headquarters in Langley, Virginia. But then the nagging doubts set in, doubts rooted in Vietnam, its horrors still in his head. When Ronald Reagan mined Nicaragua's harbor in contravention of international law, Pierre organized fellow Vietnam veterans in protest. Meanwhile, at Logicon, proposals crossing his desk filled him with increasing dread, including one to develop a system for reading another person's mind by analyzing his every blink and twitch. Pierre buttonholed the proposal's author, saying, "This is immoral, it's 1984 stuff." But the response was, "Don't worry, it's just for using on the Russians in things like arms negotiations." When Pierre shot back, "C'mon, what makes you think the CIA has any compunction about using this stuff on its own people?" the coworker clammed up. Shortly afterwards, Pierre's top secret clearance was pulled and by then he was only too glad to resign.

When I met Pierre that winter morning in a Silicon Valley hotel lobby, he was, as I say, feeling relieved and triumphant. He and his wife had cut loose not only from Logicon but from Mormonism, disagreeing with, as Pierre said, the church's "authoritarianism and weird apocalyptic nationalism." He had landed a well-paying position with Teknowledge, one of Palo Alto's leading commercial AI firms. He had been assured that his work would have no military component, and he had started a discussion group on science policy issues like Star Wars with coworkers. (Pierre's view, shared by many in his field, was that software for so complex a system as Star Wars could never be glitch free, that Ronald Reagan's space shield was therefore more likely to start a nuclear war accidentally than prevent one.) Pierre's was a story of moral high ground sought and found and peacefully occupied, and I was pleased to be able to tell it.

I never got the chance. Before I could finish the piece, Pierre invited me over to the tract home, full of yet to be unpacked boxes, he had rented for his family. Teknowledge had fired him, Pierre said, and good riddance as far as he was concerned. He hadn't much liked his coworkers ("so utilitarian . . . no consciousness about the outside world") and they hadn't much respected his skills ("One told me I had trouble forming mental models."). But how was Pierre to concentrate on programming after discovering his faith had once again been misplaced, that top minds at Teknowledge were busy courting Defense Department contracts after all? Pierre had tried with all his might to draw certain definite lines around his work, his life, his soul, but what was the use of drawing lines if your employer sneaked around erasing them, leaving you to find out later? No, Teknowledge was not a place where he could be happy, Pierre wanted me to know, and he gave me a perfect example of why.

On the day after Congress approved $100 million in military aid to Ronald Reagan's contra guerrillas in Nicaragua, Pierre Blais arrived at work early and routed an electronic message to the terminals of every Teknowledge employee. It said: "Today is a day that will live in infamy in the history of this country."

"Everybody yawned," said Pierre, so next he ventured out of his cubicle and started ranting at a coworker he knew supported contra aid. "Why don't you put your bod where your mouth is? Why don't you grab an M-16 and shoot a few people? Why don't you do that instead of retreating into your moral cowardice? Why don't you do that instead of advocating that our money be used for others to do the killing for us?" What Pierre wanted me to put in my story was this: A week after he was fired, he stood before fellow protesters and returned the Bronze Star he won in Vietnam.

There was something in Pierre's voice, the loss of cool precision as he spoke into my tape recorder on his kitchen table, that made me full of foreboding for him. Outside, his daughters were riding their bicycles in the quiet, empty street. His wife, who had not looked me in the eye when I arrived, was in a back bedroom with the door closed. What now? I asked him. His finances were shot, Pierre said, the rent past due. Homelessness was a real possibility, that and the break up of his marriage. And yet, his only work these days was for free; he was giving his time to a start-up company that swore off all military money. The company owners, headquartered in a Santa Cruz cottage, had promised to pay him when and if their product—artificial intelligence for personal computers—ever panned out. And if it didn't, well, by then he might be long gone anyway, moved, Pierre said, to New Zealand or Japan, or to "some other country where they need high tech people but are more serious about peace."

For a brief instant, a self-important worry crossed my mind, the thought that I might have egged on Pierre's search for absolute purity, making me partly responsible for ruining the comfortable California lives his wife and children no doubt had assumed would be theirs. Pierre, with his tract home payments and embattled soul, was about the age my father had been at the height of the Vietnam War, the age when my father had worn a black armband into Lockheed and a supervisor had told him to save his job by taking it off, and my father had complied so that he and my mother and we children would not have to experience what Pierre was causing his family to experience now. But, no, of course there was nothing in me that could have made Pierre come to this. All of it was deep within the man, a noble obsession surely, but one that now verged on the quixotic, one that would go on exacting a sobering price that I could not imagine his family enduring. I wondered, in fact, as I searched Pierre's agitated eyes, if he physically could survive it. In hackneyed fashion, I tried to put a name to what might be within Pierre, asking him if he had ever really gotten over what happened to him in Vietnam.

His answer was, "If so-called post-Vietnam traumatic stress syndrome is moral pain, then I have it."

I put that quote toward the end of my freelance piece, which ran in February of 1987 in the *San Francisco Examiner*'s Sunday magazine. Not long afterward, the Santa Cruz start-up company went bankrupt, its product a bust. I had by then already lost touch with Pierre Blais, had moved on to other stories. Recently when I went looking for him again I had no luck, and so I cannot say how well or badly it has gone for him, his wife, and their three daughters.

In the final spring of his presidency, my family found ourselves basking, once again, in the sun of Ronald Reagan's hometown. This time it was my younger sister, Maggie, who was graduating from the University of California at Santa Barbara with a bachelor of arts in Spanish, and it felt good to be reassembled in so lovely a place for such a happy event. So many good things had come our way over the past eight years. Deirdre and I had married and we were living in a San Francisco apartment with a panoramic view while she pursued a PhD in education at Stanford and I worked as an editor at Pacific News Service and then, better paying with benefits, the *San Francisco Examiner*. My sister Marybeth had met her future husband, a medical student, and was earning a master's degree in physical therapy. My brother Dan had graduated from the University of California at San Diego and was now in medical school at the University of California at San Francisco.

Ronald Reagan, via Lockheed paychecks, had continued to meet the bills that came to my father from these various universities. Ronald Reagan had also, by then, paid for the extensive remodeling of my parents' tract home, a project that turned out even better than expected. When the renovation was done, my parents had hired a professional landscaper to make the backyard into an arboretum of native California bushes and fruit trees, my father building a new deck behind the new family room with its billiard table and another deck off of the new dining room with its redwood ceiling and windows all around. The renewed homestead made a wonderful gathering place, and we children would come for dinner often, bringing with us friends from work or school who marveled at what a nice house, what a nice family, what a nice life. While never giving Ronald Reagan a single vote, we had prospered alongside the other aerospace families on the block who did vote for Ronald Reagan and who added on to their own tract homes their own sun rooms and libraries and professionally landscaped gardens. In the end, to have lived well in blue sky suburbia under Ronald Reagan, it did not matter whether or not you had been able to make yourself believe a single word he said.

By the spring of 1988, Ronald Reagan's claims for the powers of aerospace, his stories of a crusade to be won in the heavens, were already fading in the national imagination. Having reached an arms reduction agreement with the Evil Empire, Ronald Reagan was now asking America to renew its faith in the commonality of humankind and the peacemaking powers of two reasonably minded men willing to negotiate in good faith. The Space Shuttle program remained grounded after the *Challenger* explosion, more than two years before, had killed all seven astronauts, including the first civilian chosen for space, a grade-schoolteacher named Christa McAuliffe who had seemed to embody blue sky optimism. Ronald Reagan's Star Wars "vision of a future which offers hope" had proven a mirage, a violation of too many laws of physics and common sense, so that now even the people in charge, the Strategic Defense Initiative Office, had come to admit the sham of an impervious space shield and had quietly redirected their efforts toward creating a system that (after perhaps one hundred billion more dollars) might stop a small percentage of warheads from landing on America, but nothing, certainly, that would render nuclear weapons "impotent and obsolete." This did not surprise my father, who, from the early days of Star Wars, found consensus among his coworkers at Lockheed, a major Star Wars contractor: "Everyone I talked to was skeptical of the whole thing. Everybody I talked to figured, 'This thing will go on for years and nothing will come of it.' But nobody was turning down the contracts." In the spring of 1988 Ronald Reagan's military buildup rolled on with a momentum begun years before, when certain language, certain rewritings of myth, certain calls for renewal of faith had been necessary to get money moving from one segment of the population to another, from the undeserving to the deserving, from the losers to the winners.

So it was that one morning that spring, members of my family were sitting in a cappuccino café on State Street in Santa Barbara, having just attended Mass in the old mission, now passing time until the graduation ceremony would begin a few hours later. Like an apparition, an ugly man appeared behind my mother, a man with a matted beard and dirty clothes and a crazy air about him, a homeless man who had wandered into the café from the encampment of fellow homeless on the nearby grounds of the courthouse. He wanted some money, and I, the person in the direct line of his vision, turned my head from him, choosing not to encourage, waiting for him to move on. But he did not. He lingered and forced his way into the reverie of togetherness at the table. He insisted on saying, "I can do something none of you can. Wanna bet?"

"No, we don't," I said. For some reason I wanted to assume the role of family voice, to move to take control of the moment and thereby protect my mother and father and brother and sister from any further discomfort. "No, we don't. No, thank you."

With that, the man reached up to his right eye and removed it, holding in his outstretched hand a staring glass marble. He was laughing then with his good eye winked closed, which made the sudden hole in his face all the more dark and empty and unforgettable. He was still laughing as he turned and left us, just the crazy laugh of one of Ronald Reagan's homeless if you wanted to hear it that way. Of course, he needn't have been crazy at all to have enjoyed taunting us with the fact of deprivation in the midst of casual affluence, and our self-preserving weakness in the face of it. He could, in fact, have been a mystic who saw into the souls of nicely dressed people in cappuccino cafés, and who found a particularly interesting case in my own. He could have known that, try as I might to pretend otherwise, as a child of aerospace I had grown up favored by every Congress and presidents Democratic and Republican alike, had been designated a winner in America's militarized economy from the day I was born, had traded on that privilege all through my young adult life, which happened to coincide with the era of Ronald Reagan. As one of the favored I had been free to slip in and out of the worlds of ghetto students and Maya refugees and Caribbean dirt farmers and soot-inhaling foundry workers, free to wander that landscape of misfortune and then step away, whenever I wished, for a Zen Buddhist brunch by the Bay or a coffee with my family in a Santa Barbara café. Perhaps the homeless man with one eye saw all this and particularly enjoyed my belief that mine was a life that refused my patrimony, he laughing at so absurd a notion, as I do today.

NINE

SEEKING CONVERSION

"I was in it so deep. I was in it so deep that, well, I just didn't have the moral compass to say to myself, 'Sure you're in it deep, but get out of it now. Cut your losses.'"

My father was making one of his confessions, one of his self-lacerations in my presence that each time had the unspoken effect of bringing us closer together. As if handing me pieces to the tantalizing puzzle of himself, he had been saying such things to me for several years by the spring of 1990 when I was thirty-two and my father had been an aerospace engineer for more than three decades. Those years had made me less his child and more his friend, he and I liked to say. As friends we talked of shared interests, something he'd read or something I'd written, some absurdity of modern life we perceived alike and enjoyed having a good laugh about. The less I was his child and the more I became his friend, the tighter my father seemed to grip me in his bear hugs of greeting and the less abashedly he would say to me, "I love you, son."

"I love you, too, Dad," I would answer. And then, if my father happened to slip into one of his confessions, one of his ever more harsh appraisals of how dull and suspect had been his working life, I would love him all the more. For I accepted that self-negation as a kind of gift, coming as it did from a father who had once seemed all-powerful to his child, powerfully charismatic or powerfully fearsome depending on the moment, but an enigma self-enclosed, self-sufficient. "It gives me *supreme* satisfaction that you did not follow my example and become an engineer," my father now liked to say to me, his friend, and whenever he did, I knew he wanted to share with me some more regrets and doubts, more glimpses of what a life spent in the military contracting bureaucracy had cost his spirit.

What gave this conversation in the spring of 1990 a different color, however, was the fact that the Cold War was won, having been ended by the revolts in the Eastern Bloc. Because of this, life in America's military contracting bureaucracy seemed to be ripe for change. The nation was abuzz with expectations of a Peace Dividend, a windfall to come from the now indefensible $300 billion defense budget. Perhaps some of that Peace Dividend would pay people like my father to make blue sky technology for the new age, magnetic levitating trains and electric automobiles and a space station or two for monitoring the earth's ecology. What a difference a Peace Dividend could make, not only in the life of a nation but in the heart of a soured aerospace engineer—assuming, of course, that the aerospace industry could be, in the popular term of the day, "converted" to peaceful pursuits. That is what my father and I were discussing in the spring of 1990, the possibility of conversion.

Days before, I had given my father a taped speech by the leading evangelist of conversion, Seymour Melman, a professor of industrial engineering at Columbia University. Melman's mixture of technicalese and hellfire preaching was different from peace marchers moralizing against "merchants of death." Melman found his sins in waste and inefficiency and needless drag on the mechanism of the American economy, sins from an engineer's book, sins that went straight to my father's qualms. Melman's accusation against the military sector was that it siphoned off too much capital and brainpower from vitally productive sectors in the economy, a critique he made as early as 1965 in a book called *Our Depleted Society*, which he followed with *The Permanent War Economy* (1974), *Profits Without Production* (1983) and *The Demilitarized Society* (1988). Never in all those pages had Seymour Melman relented in his damnation of aerospace culture, its bloated bureaucracy and guaranteed profits hidden behind a curtain of black budget secrecy, sins bringing punishment upon all citizens as America's productivity rate declined and the Japanese whipped us in the commercial marketplace. And now, with the Cold War over, the lone rationale for so pernicious a culture, "national security," was evaporating.

Seymour Melman's vision of conversion looked far beyond the new products that must roll off aerospace assembly lines. The very culture within aerospace plants must be converted to the doctrines of the commercial marketplace. The United States government (which had, after all, fostered the military contractors' mindset) would force this great reformation by requiring Lockheed and its ilk to immediately create conversion committees made up of rank-and-file workers as well as management. Those committees would chart each company's new, peacefully productive future. Who should be laid off? Who should be retrained? What now should be invented and manufactured by this firm? You couldn't let stockholders and top management decide those things, Seymour Melman argued, because their interest was short-term profit, and so they would likely sell off assets and fire employees en masse and otherwise stick to making as many arms as possible, leaving America a nation weakened economically and technologically. But if ordinary workers could be given a say in their fate, they would map long-term profitable—and *productive*—new missions for their companies, transforming the shape and culture of those firms while insuring jobs for themselves in the process. The keys to their own salvation would be placed in their hands.

I found myself very much wanting to believe in this vision of a saved and reformed Lockheed, this idea that the blue sky good life I had known as a child need not end just because the market for blue sky weaponry was disappearing. I wondered if my father could himself believe, and so, after he was done hearing Seymour Melman's gospel, I decided to tape-record his reaction.

He came to me with a yellow pad full of notes in hand, saying, "Yes, I agree that Melman's ideas are going to have to be addressed. Will they come to pass without some kind of convulsion in the military-industrial complex? I think not. The first thing Melman's ideas will have to endure is an incredible, entrenched resistance by those whose careers are at stake. These are powerful men with very powerful interests in the status quo. Now someone like Melman comes along and says, 'The game's up, guys. You are obsolete.' The first thing someone is going to think of when he hears that is, 'I'm not gonna settle for that, because if I agree with what he's saying, that makes my whole *life* irrelevant.' You are going to have to literally blast these guys, blast these ideas, out of their economic foxholes.

"That's the ambivalence of it for me," said my father. "I realize I am one of those people to whom it would be announced, 'Your adult working life has been spent in a futile pursuit. You're not needed anymore.' That would be a bitter pill to swallow. But, to tell you the truth, the idea's been creeping up on me for a long time. The more I asked myself, 'What the hell is this doing for the species?' and the more I saw of dissipated energies and squandered talent, the less enamored I was with the aerospace industry.

"By the time these ideas began to dominate my thinking, though, I was in it so *deep*. I was in it so deep that, well, I just didn't have the moral compass to say to myself, 'Sure you're in it deep but get out of it *now*. Cut your losses.' "

Never had my father revealed to me so much of the burden carried within, no longer merely hinting at a career uninspiring or even unworthy of his potential, but voicing the fear that he had traveled too far along a path morally doomed. Of course, never before had there been a Cold War finished, a Peace Dividend coming. What I took from my father's latest confession was that we two suddenly shared an interest in exploring the potential for conversion, and in it, a kind of redemption.

Electronic mail from a graduate of Arizona State University, class of 1990:

> *I have a hard time with all the hype about getting the nation's children interested in math and science. I got caught up in the hype for many years. I was born in '67 and can even vaguely remember sitting on my father's lap, watching Apollo launch in '69. Ever since then, everyone promotes science, and especially AIR science! I have a degree in aerospace engineering. I want to be one of those guys you used to see in the TV ads, on the NASA mission launches, in the videos, and in the marketing the government publishes on why a student should enjoy math and science. I want to design rockets, jets, shuttles. I want to model airflow, run tests in wind-tunnels, build mock-ups. I want to be the one who says "10, 9, 8, 7, 6, . . ." But I am having a hard time finding a job in the aerospace industry simply because my expectations are too high.*

A series of vignettes will tell how my father and I fared in our search for conversion, a quest that ends, some four years later, back in the living room of my parents' home.

In the first scene it is the spring of 1990, still, and I am in the basement of Columbia University's industrial engineering department. Seymour Melman bursts through the door of his tiny office late for our appointment, yanking off his cap to reveal a snowy thatch, landing in his chair with a wild grin. Melman is rejoicing at a recent *New York Times* editorial telling President Bush to get busy cutting the defense budget in half by the year 2000. After that ran, a *Times* editor invited the professor over to hear his gospel of conversion, which Melman has delivered just this day. "It's the missing link!" Melman is fairly shouting at me. "The editor owned up that what's left open is the conversion issue! Conversion is the missing link! What's the use of talking about budget reduction, and what you're going to spend it on, if these bunnies in between are scared to death!"

By "these bunnies in between," Melman means the politicians who want to vote a Peace Dividend, but not if it costs too many jobs in their districts. The solution is his conversion approach, Melman promised the editor as he is now promising me. In fact, Congress could easily surpass the cuts called for by the *Times*, saving the country well over a trillion defense dollars by the year 2000. And, in a twist on Ronald Reagan's supply-side economics, Melman calculates that reinvesting the savings into infrastructure, education, and R&D would create more than enough wealth and jobs and tax revenue to erase the deficit within a decade.

With the Red Menace kaput, Melman is busy these days evangelizing peace groups and labor unions and Rotary clubs, drawing enthusiastic applause from all. A conference of mayors has given him an ovation and a unanimous resolution calling for conversion. Seymour Melman has even preached to Bryant Gumbel on the *Today* show. The logic is inescapable. Surely anyone who stands to gain from government spending on butter over guns—teachers, librarians, road workers, police, medical researchers, the homeless—represents a natural constituency for Melman's plan to phase out the military economy. Congress cannot for long ignore such mounting political pressure and so it is only a matter of time before the bill in Congress that Seymour Melman has helped frame will be passed into law and the great reformation can begin.

That is why Seymour Melman is fairly shouting at me. He is gripped with the fever of a prophet vindicated by unfolding events. "I mean, you can demonstrate that the Soviet military machine has turned into a ladies' afternoon tea club and that would have no effect on the pressures for keeping the bases, keeping the contractors, and keeping the military labs funded. It's clear the conversion thing is the missing link between budget reduction and the Peace Dividend!"

A few days later, I am in Washington, D.C., come to see how conversion is proceeding on the Hill. Dick Greenwood, special assistant to the president of the International Association of Machinists and Aerospace Workers, is smoking a cigarette in his office and telling me why, way back in 1978, he helped draft the very first Seymour Melman inspired conversion bill, a bill submitted in roughly the same form every year since and rejected every time. "At that time, everybody was telling us that one third of our membership was involved in military production. Part of our strategy was to try and resist, somehow, marching arm in arm to Capitol Hill with our employers every time a defense contract was threatened. We had to have a program that would permit us to take an enlightened view of foreign and military policy, that could be supported by our people. We weren't asking for blanket cuts; let the rubble fall where it may. We were asking for a reinvestment strategy to put back into the holes that would be created by those defense cuts."

And that, Greenwood says, means serious money for retraining and relocating workers who fall through the holes. Democratic Representative Ted Weiss of New York has submitted the Melman bill again this year and it is the only one with serious money for retraining and relocation, which is why Dick Greenwood's union is backing it strongly.

But what about Seymour Melman's grand plan? I ask. What about workplace democracy, the conversion committees to be mandated inside of every big firm? Greenwood gives a raspy laugh and tells me he doubts management would pay any attention to those committees. "Their bottom line is, maximize profits." He takes another drag, his laugh raspier and rueful. "And if they lose a contract, they'll just strip the company."

My pilgrimage to the Hill continues with a visit with Representative Weiss, a gentle, thin-faced man with large ears and an endearingly bad haircut. Having spent the previous two days speaking—off the record, always off the record—with various congressional aides close to the action, I am telling Ted Weiss that his conversion bill is this year, like every year before, a dead letter. He listens patiently as I tick off the camps opposed to conversion: the hawks who want the defense budget preserved; the deficit hawks who want defense cuts without a Peace Dividend; the Peace Dividend liberals who want it spent on social programs but not on new lives for affluent aerospace engineers. Even the pro-conversion camp is split, most members rolling their eyes at Seymour Melman's notion of forcing contractors to include workers in planning new business. Political action committees for weapons makers have spent two million dollars on key congressional seats in the latest election and not coincidentally, as one aide puts it, "There are two kinds of Democrats. Those who are willing to piss off contractors and those who aren't. And that split is really paralyzing the party."

Ted Weiss listens without disagreement as he stares out his window. He says, "This came out at the Democratic caucus the other day: Of the top one hundred contractors, sixty are under indictment or under investigation. So I don't know what the merit is in kowtowing to contractors. They are not our best citizens."

It is getting toward the end of my spring-of-1990 Washington visit. A bulldogesque, Democratic representative named George

J. Hochbrueckner is taking me through a dizzying show-and-tell. His prop is a poster-sized chart kept handy by his desk, a tally of various aircraft meant to illustrate that there are not enough Grumman airplanes in the world, and in particular not nearly enough Grumman F-14D fighter jets.

George J. Hochbrueckner happens to represent the Long Island home of Grumman Corporation's big complex, happens also to be a former Grumman engineer who won the last election with 50.8 percent of the vote in a district that is largely Republican. After winning, he joined with the rest of the Long Island congressional delegation, conservatives and liberals alike, in the relentless quest to restore a defense budget item, the Grumman F-14D fighter jet, which even the Republican Secretary of Defense, a hawk's hawk, said America didn't need. America got eighteen of them, anyway, costing the taxpayers one and a half billion dollars.

"If this was strictly a pork-barrel thing," the congressman from Grumman is now solemnly assuring me, "there is no way we could win." For proof I need only examine his chart. "Based on the Navy's new, revised numbers, [u]sing their own numbers, we're fifty-six [F-14s] short. So the reason we went after F-14 is we could make a legitimate case the Navy would be short of aircraft." The Navy, in other words, needed more planes because the Navy said so, and that's why this was no pork-barrel thing.

Now Hochbrueckner is waving his hand over the entire chart, noting that the Secretary of Defense has proposed axing every plane on it. "And that's a budget that wipes out Grumman. And then a couple of years from now, if we said, 'Oh, my god, look how short we are on these airplanes,' Grumman won't be there to build them. So last year's effort was a genuine one, in that we could make the case that national defense required these aircraft."

What confounds me is that George J. Hochbrueckner has signed his name to every conversion bill in the House. He wants, for example, millions of federal dollars devoted to building a high speed, magnetic-levitation train like the one Japan is developing.

"Perfect" for the job, says George J. Hochbrueckner, is Grumman Corporation. My head spins with the Alice in Wonderland conundrum that is George J. Hochbrueckner's "conversion": Weapons makers will make the weapons they've been making, even when the leading warrior in the land says we don't need them, even when the same George J. Hochbrueckner wants more tax money poured into nonweapons projects to be carried out by the same weapons makers.

Through the haze of my confusion, I hear, "We had to do this one, the F-14, because this is the bread-and-butter aircraft for Grumman. That's the one that makes the money. Clearly I was successful at making the case that we're short of these, and therefore, from a national defense point of view, we should buy them. Now what this does for the company is buy them another two years. The last of these aircraft doesn't pop out of the pipeline until April of 1993. One of the advantages of winning this was meeting the defense need, but also buying Grumman some time, so programs like Ted Weiss is pushing can be implemented . . ."

My father's words keep intruding on my concentration. *You are going to have to literally blast these guys out of their economic foxholes* . . . When I tune in again, George J. Hochbrueckner has worked himself into a lather of pride in the service he has rendered.

"At this point Grumman is geared for success. They're saying, Okay, we've got nineteen F-14Ds we're building now, the eighteen we just put in last year—they'll be building those things until April of 1993 and maybe longer. There may be more F-14s in the future, there may be a Tomcat 21, which is Grumman's version of a replacement for the F-14. So I think Grumman's view is, 'We'll be building aircraft for many years to come . . .' "

E-mail from an engineer who describes himself as working "in a little corner of the Space Shuttle program in Houston": *I don't see a future in my job. Many people in my area have gone back to school, usually for MBAs. Others have either left or been laid off but no one ever leaves to go to another aerospace job. If the picture I'm painting hasn't told you already, morale seems to always be bottoming out. I'm not a Cold War veteran. I joined the program in 1988 just before the first flight after the* Challenger *accident. In college I was fed a steady diet of Mars missions, space stations, advanced fighter technology, etc. and I ate it up. I came down here ready to jump into the race. I found bitter disappointment at small-minded bureaucrats who loved to tell me great old war stories about the Apollo program. It feels like "they" told "us" that we were needed and the space program was going somewhere, but "they" were full of shit.*

It is March of 1992 and my father, mother, and I have driven to San Jose's redeveloped downtown to see Silicon Valley's new shrine to itself, a museum called The Tech. We are able to walk through a microchip clean room, program a robot arm to arrange toy blocks, appreciate a colorful laser pattern, inspect an ultralight, superfast bicycle. Everything is user-friendly and human scaled. The only nods to aerospace are a display about how the screwed-up Hubble Space Telescope might eventually be fixed, and a television screen flashing images of Mars that make the planet's surface look much like the outskirts of Yuma, Arizona (the pictures were gathered by a Viking satellite fifteen years before). There are no missiles, even though Lockheed Missiles and Space Company is still the area's largest single employer and hasn't stopped production of the Trident II. There is none of the spectacularly murderous technology that just a year before had mesmerized the nation watching Desert Storm on CNN, no smart bombs or stealth fighters, parts for which are made locally. The most unsettling item I find is a notice that McDonald's researchers are at work on a fully robotized fast food kitchen. At

The Tech, technology is ingenuity and play, having nothing to do with military budgets.

The sense of denial pervading The Tech seems to inhabit the national psyche. As I had foreseen, the limp conversion legislation eventually passed by Congress requires no reformation of blue sky corporate culture. Like the magnetic-levitation train and the other big ticket, "peaceful" missions for aerospace once bandied about, the Peace Dividend never materialized. In servitude to pork-barrel militarism, Congress built a fire wall around the defense budget, limiting cuts to a scant 2 percent per year.

The George J. Hochbrueckner version of conversion seems to be winning out after all. Business as usual, that is, and business wherever it can be found. About the time my father, mother, and I are wandering The Tech's exhibits of fun, leaders of six of the nation's largest defense contractors are penning a letter urging President Bush to quickly endorse the sale of seventy-five F-15 planes to Saudi Arabia, their argument having less to do with global security, more to do with the economic prospects for families like mine. "It would rapidly inject five billion dollars into the economy, reduce the U.S. trade deficit, and sustain 40,000 aerospace jobs and a corresponding number of jobs in the non-aerospace sector of the economy—all at no cost to the U.S. taxpayer."

To my blue sky family, I must admit, The Tech is rather boring. No appeals to awe, to the conquering of anyone or anything. The message I take away is that Silicon Valley, like much of America, would rather not dwell on its military self now that a compelling enemy is lacking, would rather pretend that the weapons industry and the people whose livelihoods depend on it are yesterday's news. After forty minutes poking around The Tech, my father, mother, and I are on the freeway leading away from downtown and back to the suburbs.

> E-mail from a Lockheed engineer with fifteen years' experience: *When I graduated college aerospace was beginning to spool up for the go-go 80s. I got sucked in*

because it was the hot field. I have been trying to get out of aerospace ever since, but have never been able to cross that line, and now am labeled and categorized solely on the basis of my aerospace background (and these are not good labels or categories). The way things are going, I may soon get the opportunity (i.e., the boot) to start a new career at mid-life. My biggest reservation with aerospace is the fact that there are no more big problems to be solved or products to be invented. Most people are bored with space, rockets, airplanes, etc.

It is February of 1993 and a new prophet of conversion has come to my hometown. Despite the glacial rate of its shrinkage, a smaller defense budget has begun to show its effects over the past year as headlines tell of tens of thousands of aerospace workers laid off across the country. Now, everywhere he goes in Silicon Valley, Bill Clinton is cheered for his revival of the Peace Dividend promise, his vow to spend what it takes to train blue sky workers to do something else, and to shift billions from weapons projects to more fashionable goals like a "clean car" and "gigabit" computer network for all.

It is fitting that Bill Clinton, who wants to be a new type of Democrat, finds a role for government in creating new gizmos intended for private utility and enjoyment and profit. There is nothing in his gospel to remind one of Clinton's claimed hero, John Kennedy, who dedicated government to the task of building and populating the Cold War technocracy, who declared that "We choose to go to the moon in this decade . . . because that goal will serve to organize and measure the best of our energies and skills."

What will organize and measure the best of our energies and skills from here on out, Clinton preaches instead, are the uncontrollably escalating rigors of global capitalism. You will have eight careers in your lifetime, he is fond of telling the American public, sounding a concept foreign to my father and his generation of organization men, but describing a way of technological life pioneered by the tribe of Woz. The optimism that Bill Clinton exudes says the American can-do spirit now proves itself in how cheerfully we train and retrain for the eight careers the marketplace demands of us. What will come of it all—what industries, what products—are no business of the government, are nothing like a moon rocket. It is each *individual,* in Bill Clinton's gospel, who is the ongoing conversion project. At best, we are promised a government arm around our shoulders as we adapt to our kaleidoscopic futures.

During one visit to a Silicon Valley company, the young, modern president is paid tribute by technology used to create the special effects in *Terminator II.* As Bill Clinton watches, faces on a video screen "morph" from George Washington to Thomas Jefferson to Abraham Lincoln to Franklin Roosevelt to John Kennedy to Bill Clinton. Ronald Reagan, who sent more military dollars per capita to Silicon Valley than anywhere else, is left out of the morphing as the audience claps and Bill Clinton offers one of his ingratiating, lip-biting smiles.

> E-mail from a twenty-year veteran at one of the largest military aircraft manufacturers: *Either you laugh or you cry. People are leaving (voluntarily) at a rate 4× the "historical rate." This is a company that runs on fear, in terms of personnel management issues. It's kind of interesting: Working-level people feel (rightly) that they're doing a good job and they needn't take a back seat to anyone. The personnel policies, however, almost seem designed with the intent of driving down morale so that people will quit.*

It is April of 1994 and once again I am seeking conversion upon the Hill. This day I am in the office of George Brown, Democratic Representative from Southern California. He is a genial fellow with a grandfatherly face and one of the more intriguing résumés in Congress. He studied physics before the atom bomb was invented. He was the first member of Congress to oppose the Vietnam War. He politicked early and hard for projects ranging from renewable energy and ozone layer protection to the space station and the superconducting supercollider. He is today the Chair of the House Committee on Science, Space and Technology at a moment when the Congress and White House are both under Democratic control. He is a man, in short, who has long hungered for the opportunity now at hand, the chance to lead in the greening of a new, federally funded technoscience agenda for America.

He is beginning to realize, the day I visit him, that his chance will never arrive. Five years after the fall of the Berlin Wall, the Pentagon still funds about half of all federally sponsored research and development. Meanwhile, the entire federal R&D budget is dropping just when many corporations are downsizing their labs, causing some experts to predict a glut of underemployed PhDs soon. Thousands of physicists are already scrambling for work now that their superconducting supercollider is a half-finished ruin in the Texas desert, Congress having pulled the plug on that monument to Big Science.

"The nation is never united so much as when it is threatened by something," George Brown muses, harking back to those misty olden days, the Cold War. "Once that pressure is over, then things tend to ease up and you no longer have a sense of a society devoted to a common goal. Religion doesn't provide it. Political leadership doesn't provide it. Writers don't provide it. Of course when you have this image of how great you are and how valiant your battle against the enemy is—in this case, the Communists—it provides a sort of negative unity. You're not sure of what you are trying to achieve. You know you don't want the bad guys to overcome you, and that's what held us together. Now we don't even have that.

"No sense of unity. No sense of unity," says George Brown,

summing up his fading hopes for a nation's ability to dream a common future, and for his government's ability to coherently plan for and invest in the technoscience to help bring about that future. "If I am hearing you right," I say, "you are close to despairing that a political coalition can ever be built that will be strong enough to support the kind of government-led science and technological development that we saw during the Cold War."

"I am becoming somewhat pessimistic about that happening," George Brown answers.

As he does, neither of us has an inkling that he is soon to lose his leadership of the House Committee on Science, Space and Technology. That autumn, the Republicans, riding a wave of anger at a federal government that no longer inspires, will sweep into power in both houses of Congress. When looking to the sky for a new science and technology agenda, the Republicans will find it easier to imagine a nuclear attack than a stripped ozone layer, and so they will push for a 25 percent increase in Star Wars funding and try to stall a ban on chloroflouride production. The two Republican members of Congress authoring the bills against the CFC ban will be named, aptly, Doolittle (John) and DeLay (Tom). The leader of the Republican charge into power will be Newt Gingrich. A basher of government welfare and a self-declared true believer in the privatized free market, Newt Gingrich will, nevertheless, also be a stalwart friend to the aerospace industry and higher military budgets, coming as he does from the Atlanta suburb that contains Lockheed's largest complex.

Months after my visit and shortly after the Republican victory, I will dial George Brown's number only to hear a taped message from Skip Stiles, his aide on the Committee for Science, Space and Technology. Making reference to the Borg, a race of lockstep, soulless automatons featured on *Star Trek*, the aide's voice will flatly inform: "We are Skip. As with all Democrats we have been assimilated according to the post-election plan. Leave a message at the tone. Or hit zero to speak to another unit in central processing. Thank you."

Late in the summer of 1994, a pleasant, kind-faced woman named Carol Alonso is telling me how it used to go when she would get one of her creative flashes. It would come, sometimes, at home in the evening, as she gazed into the fire. The idea would take shape the next morning as she showered, made breakfast for her husband and two children, pulled her Porsche out of the driveway of her split-level home in the Northern California suburb of Orinda. At the office she'd try the idea on her closest workmates. If they liked it, she'd create some computer models, deliver some briefings, submit her idea to rigorous scrutiny by the best in the business. Every step would grow more and more competitive and Carol Alonso had "seen grown men cry" as they watched their ideas "picked to pieces." Still, when one of hers made it through, the pain all seemed worthwhile, for Carol Alonso would be handed the prize she and her peers had spent their lives chasing. She would be given America's next nuclear explosion.

Then would follow the heady days leading up to an underground test, Carol Alonso directing her team toward the "shot" deadline a year away. If this meant Dad and the kids saw less of Mom, the Alonsos, a thoroughly modern family, coped. After all, they had been through this many times before. And, too, they had something to look forward to. Once Mom's creative flash did become a thermonuclear reaction causing the Nevada desert to rumble with the unleashed force of 150,000 tons of TNT, once Mom's bomb *did* go off, the family would celebrate together, unwind, escape. They'd all take a nice vacation.

"We sailed together in the kingdom of Tonga. We hiked in the Himalayas. We sailed in the Caribbean several times. We went to Tahiti and sailed around it in a chartered boat," says Carol Alonso, eyes alight with the memories. "We had wonderful times together."

Wonderful times together for Carol Alonso and her fellow nuclear weapons designers are rumored to be nearing an end the day she and I talk at her workplace, the Lawrence Livermore National Laboratory. Treaties oblige the United States to cut its warhead stockpile by nearly half within the decade. Gone is the heyday that saw up to four shots a month, more than 1,000 explosions since 1945. President Clinton's renewed moratorium means there hasn't been a single underground test for three years and the world is moving toward a comprehensive test ban. Some members of Congress, Republican as well as Democrat, are urging that Lawrence Livermore be shut down and its work be shifted to its sister lab in Los Alamos. Weaponeers tell me the mood of the place is souring as they see slipping away their reason for being. "We're subcritical now," one old-timer dourly confides, fearing there are too few left with the know-how to carry on the tradition of nuclear bomb making.

In reading about the crisis of purpose in America's nuclear laboratories, I see that MIT anthropologist Hugh Gusterson and I independently have come to share a metaphorical perspective. He has studied nuclear weaponeers for years and finds them to be a tribal society whose central ritual was the underground test that Carol Alonso so fondly remembers. For elders, the test built stature and gave them "a feeling of power over their weapons." For young weapons designers, the test was "a rite of passage" into the tribal inner circle. But now their central ritual is banned and so the nostalgia and fretting I encounter at Lawrence Livermore should be no surprise.

What surprises me is where hope has been found to live. The nuclear weaponeers see a way to keep their central ritual, if only in "virtual" form. The labs and their allies are campaigning for a new generation of technology that will allow nuclear explosions to be simulated in the laboratory so that the design—and testing—of nuclear weapons concepts can continue apace. The centerpiece of this program would be a $1.8 billion superlaser to be located at Lawrence Livermore. It would be the largest single military project ever built at the lab, and a magnet for more weapons talent, money, and work. Critics argue that "virtual" testing is an expensive boondoggle that will make a mockery of test ban efforts and promote nuclear proliferation. But the weaponeers I speak with at Lawrence Livermore are hopeful that President Clinton will give the go ahead on the superlaser, hopeful that this will stave off the extinction of their culture. (Their prayers for rejuvenation will be answered two months later as the superlaser receives White House approval.)

On the day I visit, one of the most upbeat Lawrence Livermore denizens is a bearded baby boomer named Kent Johnson, who tells me he was heartened to see that a recent job posting for a nuclear weapons designer drew forty applicants. If the superlaser comes through, he suggests, think of all the fresh blood the tribe might initiate. "I really believe we can recruit people," he says, "if we have the money." The fact that the nuclear weapons stockpile is officially said to be 8,000 warheads too full at the moment when he is telling me this does not seem to matter to Kent Johnson, who remains very much interested in the design and testing of nuclear weapons, who expresses no interest in finding other work. Kent Johnson snaps my wandering thoughts back to our conversation by asking, "Your dad. What's his name?" He is intrigued that I have mentioned my father's career as a Lockheed Missiles and Space Company engineer. "I've worked with a lot of Lockheed folks," says the nuclear weapon designer. "I might have worked with your father on a project or two."

"No, I'm sure you haven't," I say, though of course I can't be sure, so ultimately intertwined are the various projects of the extended tribe, so translatable, across the various subcultures within the extended tribe, are certain beliefs and symbols and language and ritual, and the instinctive desire to preserve all this.

The electronic mail is beginning to trickle onto my computer screen in the autumn of 1994. I have sent a message to a few sites on the Web where aerospace people converse. I have asked, "How does it feel to be in aerospace today?" and every time I check in, a few more answers await me. Today there is this from a twenty-nine-year-old aerospace engineer:

> *My first job was with McDonnell Space Systems Co. on the Space Station program in Houston. After a few years I quit to take a job with a small US Navy contractor in Virginia. Layoffs were always hanging over everyone's head, especially at McDonnell Douglas, where I survived three significant ones (at least 50 people). The trend there was to first lay off the "dead wood." Usually these people were capable of doing a good job but had a bad attitude or were just plain lazy. The next to go were those unlucky enough to be in the wrong place at the wrong time. They came from every level—upper, middle, and lower management and the engineers, aides and secretaries. You just couldn't accuse McDonnell Douglas Space Systems Co. of discrimination!*
>
> *The other company I worked for was diametrically the opposite of MDSSC in several ways: 1. Everybody knew everybody. 2. Lame benefits (small companies just don't have the money). 3. Poor management (incompetent). 4. Quality people were lacking (you get what you pay for). 5. People just weren't as nice. Well, I recently quit that job and am at the University of Kansas working on an M.S. in aerospace engineering with an emphasis on aircraft design. Honestly, and I know this sounds silly, you need to be an optimist in this industry!!*

There is this as well from a senior aerospace specialist at AlliedSignal Engines who asks, like nearly all the others, that I not reveal his identity for fear that being seen to complain might further dim his worsening prospects:

During my 14+ years here—formerly Allied-Signal Engines, formerly Propulsion Engine Division, formerly Garrett Engine Division, formerly Garrett Turbine Engine Division, formerly Airesearch, well, you get the idea—and GE-Evendale, I have watched the aerospace business become a truly depressing place. Sadly, this company and many like it are being run only for a few people in the boardroom and Wall Street. Yes, there is no more Soviet Union, and yes other economic pressures have forced some contraction in employment, but consider that this company is being hit with layoffs every quarter if profits aren't on target, even though in its 50+ year history there has never been a losing quarter. No matter to the CEO; it isn't enough that we're profitable, it's that we're not profitable enough. I've watched more than half the engineering workforce leave, either voluntarily or not, during the past 4 years and the end is still not in sight.

The saddest thing has been the change I've witnessed in the workplace over all these years. In days of old, there was an incredible amount of camaraderie and high spirits. At the end of the day or the week, one socialized with his/her colleagues as they were truly friends as well as coworkers. We were happy to put forth whatever effort was needed to get the job done. All that has changed. It's turned into a rat race. We have a joke here—it's no longer a career, it's just a job. Most of us are looking for other positions, so we're digging in until they lay us off or something better turns up.

The next day a graduate student at the Colorado Center for Astrodynamics Research modems this to me:

In 1987, when I began college there were articles and articles about how many aerospace engineers were needed going into the 21st century. Now, still being a student, these articles are nowhere to be found.

The tone of my e-mail matched the voices I was hearing, over the years, whenever I spoke with aerospace people who had been laid off. The tone resided in the narrow place between guarded hope and disgusted resolve. Few openly displayed the vitriol of one Lockheed engineer, dumped three years short of pensioned retirement, who told me that, as his supervisor gave him the news, he was thinking to himself, "I'd like to kick you in the balls." (And who continued to feel this toward Lockheed management: "I hope they all get boils under their armpits.")

More often they sounded like forty-three-year-old Chuck Goslin when he matter-of-factly stated, "I was downsized." Chuck Goslin, like his two brothers, his father, and his father's father, worked for the Grumman Corporation on Long Island. Grumman was the only job his father ever had, a job that started in 1945 and ended in 1992, and so he had known the company in a different time, when "it used to be you knew everybody and there was a lot of teamwork and you were proud of Grumman." When his father had worked on the F-14 fighter jet and the lunar landing module that Neil Armstrong stepped from so famously, "It was a hometown pride sort of thing. You had a sense the *region* did this. That's a good feeling. But the last five years have been disheartening because it all had become so competitive. We were bidding on things we knew we were never going to get. Why bid? Just to keep us busy."

A manufacturing engineer, Chuck Goslin worked with 25,000 others at the Bethpage complex at the height of the Reagan buildup. But his company lost a key space station project to Boeing in the mid-1980s, began shifting electronics and manufacturing work to plants in Florida and Louisiana, and finally submitted to a takeover by Northrop Corporation. By 1994, Grumman had laid off four fifths of its Long Island people. Chuck Goslin was handed his pink slip one morning by a manager who told him, "I have good news and bad news. The bad news is I can't keep you. The good news is you're the last person I'm laying off this year."

He was given fifteen minutes to clear out. "They were worried about sabotage, which kinda made sense. And there was a cut-off. If you'd been there twenty years, you got two weeks' notice and a nice severance package. Otherwise, fifteen minutes' notice, pack up your stuff, sign these papers and nice knowin' ya. That's what happened to me. I walked out the door, went home, called the missus up at her work and said, 'Well, I didn't make it another week.' It was a beautiful day, I remember. I drove by my dad's house. 'Hey, that's it,' I told him. 'I'm lookin' now. I'm in the search mode.' "

Chuck Goslin's search led him to evangelists of conversion. For years the local conversion activists had been pushing to diversify the Long Island economy away from Pentagon dependence. They had been crying their warnings even as George J. Hochbrueckner was assuring me that "Grumman's view is, 'We'll be building aircraft for many years to come' " because "at this point Grumman is geared for success." But now there would never be another Grumman aircraft built on Long Island, and George J. Hochbrueckner was making defensive criticisms of Grumman management in the newspaper, and Chuck Goslin had found work at the local conversion office as a career adviser to others no longer employed by Grumman. "Most of them are bitter," he said. "Angry because they haven't been able to rationalize what's happening. They get no help from Grumman whatsoever."

What Chuck Goslin could not rationalize is why conversion was such an impossible dream for a company like Grumman. Why did the region's largest supplier of high-skill jobs have to be gobbled up and emptied out? Why didn't Grumman, for example, ever get a chance to build trains or buses? "Airplanes," said Chuck Goslin. "Cut off the wings, cut off the tail, whuddaya got?"

Instead, near as Chuck Goslin could tell, here is what happened. "Grumman liked to call itself The Family Corporation. But the decisions weren't made in a family sort of way. It was done with numbers. The top forty or fifty people kept getting stock options as the company lost money and people lost jobs. Now, just today, Grumman Northrop announces its fourth quarter earnings are a record high. And two weeks before, they cut 3,500 workers. Boom!"

It is September of 1994 and my father, having handed me another beer from the refrigerator, is winning his fifth straight game of pool. "What can I do? You're invincible on your own table," I say as the eight ball banks in and my father gives one of those satisfied *shtocks* I remember from childhood. He is smiling. "I've got something you'll be interested in," he says, popping into the VCR a videotape with the familiar Lockheed star logo on its label. "The latest propaganda."

This tape, my father says, was handed out earlier in the summer by Lockheed Missiles and Space Company. Every employee was instructed to watch it at home to learn why half the workforce had been laid off in the past four years, and what those who remain might expect in the future. For fifteen minutes the CEO and a handful of other senior managers flash interchangeably onto the screen, their heads and eyes barely moving as they speak of "prioritizing," of "investing smartly" in "core competencies," and of the "need to be prepared personally for change." Woven through their declarations is an anonymous, friendly female voice who lays out the grim facts.

"The last three years or so have been among the most trying the LMSC employees have ever experienced. . . . Thousands of our colleagues have lost their jobs. There are still many employees here now who are not confident they'll be here a year from now. . . ." She revisits history to provide context. "It's 1991 and the Soviet economy has crumbled. The Cold War, perhaps more a war of economics rather than military might, has ended with the United States the apparent winner. But we as a nation have paid quite a high price for that victory. The losses to the defense industry over the last several years are staggering. Total sales for the defense industry were down 14 billion dollars in 1993. That's a 10 percent drop in one year alone, the biggest drop ever in the industry. . . . From 1989 through 1992 the aerospace industry eliminated 291,000 jobs, and then in 1993 another 131,000 jobs. This left the aerospace workforce total at 909,000, the first time in more than fifteen years the aerospace work-force had dipped below one million. What's in store for 1994?"

What is in store, the viewer is told, is more of the same. And all of this, the various talking heads of management go on to say, has taken them quite by surprise. After the Cold War ended, the downturn in "defense procurement caught most of the defense contractors unaware," says one. "I think in particular Lockheed was," points out another. Indeed, as late as 1992, three years after the fall of the Berlin Wall, the leading minds of Lockheed Missiles and Space Company issued a plan called "Vision 2000," which projected a 50 percent *growth* in the sales of Lockheed missiles, satellites, and such by the end of the decade. "Hardly less than twelve months later," another manager admits, "that really appeared optimistic."

My father offers a laughing translation. "He means: We didn't know what the hell we were talking about! *Still* don't know what we're talking about!"

Now the woman's voice is saying that things really turned sour two years ago. About the time Bill Clinton was elected, more projects were pulled, including the "cancellation of more than half of our classified business."

A manager's talking head appears. "There's been almost a vengeance against defense spending now."

"Democrats!" my father interprets with snicker. "Damn them all."

"Clearly," says the female voice of Lockheed, "something had to be done and it had to be done fast." What was done is that a number of "task forces" made up of "the best and brightest employees" were told to find ways of saving the company $46 million a year, ways that included the trimming of benefits and, unavoidably, the further firing of fellow employees. This, says the female voice of Lockheed, was "excruciatingly difficult and, at times, agonizing." Sadly she foresees more "trying" months ahead "as we continue to grapple with the volatile defense and commercial environment."

What the talking heads of management are wanting to say to everyone watching this tape is that they do have a new plan for Lockheed. The company will maintain its present size by trying to sell satellites and other products to commercial buyers, cutting the current 99 percent dependence on government to 70 or 80 percent. The company might also consider a merger with some competitor, says a talking head of management. As he does, he probably knows that Lockheed will in two months merge with Martin-Marietta, which will create the world's largest aerospace firm while causing the elimination of 19,000 jobs in twelve states and boosting the price of Lockheed stock. But he does not let on to these specifics. He says only that the way of things is that "the strong will survive and the weak will be consolidated into the strong."

"And finally," says the female voice of Lockheed, "we asked [the company's leaders]: If you could make one statement to the people of LMSC, what would that be?"

"Good-bye," says my father.

"This is such a rapidly changing world," comes the answer from a talking head of management. "Ah, we've had huge technological gains in the past but, ah, they are going to continue into the future at an accelerated pace and, ah, individuals owe it to themselves, they've got to continue education, continue improving their skills that are transferable not only among the company but ah, ah, you know, throughout the work environment . . ."

"Not a hint," says my father when the tape has ended, "not one syllable of how the company is going to help even the chosen and gifted employees enhance their skills. If you make yourself attractive to the company, we might keep you. And the way we're going to maintain our present size is to be successful in a line of business we've never been successful in before, and run by the same team of senior managers who have confessed that they were completely taken by surprise that, with the end of the Cold War, the defense budget went *down!*

"So this all makes a hell of a lot of sense to the poor befuddled employee. In the privacy of his own home, where he can't challenge the speaker, he listens to this monologue that tells him he's up a stump."

My father's cynicism is infectious and I am chortling along with him, two friends perceiving together one more absurdity of modern life. What a laughable dream now seems the notion of aerospace plants shifted to peaceful pursuits as Seymour Melman had once prophesied it, an orderly process insuring "productive" jobs for the maximum number of blue sky workers. Melman the industrial engineer had looked upon the aerospace industry as a flawed machine awaiting modification by perfectly rational criteria. He had failed to see in aerospace a national enterprise assembled and fueled by myth, a myth that would have to be replaced by an equally powerful myth before any true reformation could occur.

The myth was as Harry Truman's Air Policy Commission had capsulized it just after the Second World War: that the next war forever looms and when it arrives we shall prevail in the skies. The myth was that "self-preservation comes ahead of the economy" and so aerospace must not be confused with more mundane endeavors such as the creation of electric cars or magnetic-levitating trains or other machines that might help America perform more efficiently.

The myth that dawned with the Cold War has proven itself more durable than the Cold War, so that rising U.S. defense spending is now officially said to be calibrated for the fighting of two Desert Storms at once in separate corners of the globe, B-2 bombers are purchased by a Congress for a Pentagon that does not ask for them, and the same Congress calls for a 30 percent drop in spending for civilian research and development by the year 2002. This long after the end of the Cold War, we can see how it will be. Even as the "consolidated" aerospace giants are staffed by a dwindling number of blue sky workers with fewer years, less benefits, lowered expectations, the blue sky culture will remain unreformed, stoic in its belief that self-preservation comes ahead of the economy. Conversion is an act of imagination and springs from a pressing desire to lose the old habits, to invent a new self. Desire was lacking within the boardrooms of aerospace companies; imagination was not to be found in the nation's leaders who might have framed new myths to sustain a frightened tribe.

The word "conversion" has long ago disappeared from pages of newspapers and magazines, and my father and I don't much mention it anymore either. The price of Lockheed stock, like defense stocks in general, has remained strong throughout. The architects of defense mergers have emerged rich (talking heads on my father's videotape will be among the Lockheed and Martin-Marietta executives to reward themselves with $92 million, $31 million of it to come directly from taxpayers). Nowadays, rather than "conversion," the words used to describe the transformation of the aerospace industry are the words used to describe employee purgings in other industries: Mergers. Buyouts. Rightsizing. Re-engineering.

"The people I talked to after this tape had been distributed were horrified," says my father. "They were very depressed. These were early to midcareer people in their thirties and forties with families and mortgages and skills and energy. And they had an interest in staying with the business, because *they were in it so deep.*"

My father, who is still laughing bleakly and shaking his head as the tape finishes rewinding, will be spared the horrified depression of his friends, however, for he is one who finally did achieve conversion (of a personal kind, at least). Two years before, Lockheed Missiles and Space Company had invited my father to take early retirement. He had accepted. Get out of it now, my father told himself. Cut your losses.

TEN

OUR LADY OF IRONY

Queen of Apostles has a new Jacuzzi. You can't miss it there just to the left of the altar—waist-high concrete tub full of chlorinated, heated holy water; small fountains burbling at four corners; bottom inlaid with tile crosses; stainless steel filtration vents. State-of-the-art baptismal fonts like this don't come cheap—high five figures—but most of the flock cheerfully paid. The look of their church was tired—early Perry Mason set, cinderblock and river rock and knotty pine. It needed . . . *pizzazz;* time to redecorate! Besides, as He Himself put it: Water is Life. Except when it's not. Like when some little kid falls unnoticed into one of the neighborhood's swimming pools. Which explains why, between baptism dunkings, a perfectly clear Plexiglas lid, resistant to all potential lawsuits, seals off the surface of this most sacred of spas. . . .

There. See what I've done? I return to the church where my mother still worships, where she still asks her saints to be good to a son in his thirties, and these are the whispers that swarm my head, scribble themselves in my notebook. If you find them funny or blasphemous or both, imagine them a different way. Read them as a prayer, a furtive little prayer by one who long ago abandoned his Catholicism to become an Ironic Fundamentalist.

We make, I suspect, an all too common set, my mother and me at Mass this Sunday in the middle of Silicon Valley in the middle of the 1990s. She is the mother who thanks God for her good life in the suburbs, and I am the son whose good life in the suburbs convinced me I did not need a God. She is the mother who offered her children Catholicism as a connection to heaven, and I am the son who placed his faith, instead, in ironic detachment.

I could offer the usual explanations. I could blame the ironic reflex in me, so quick and caustic, on the false promise of blue sky suburbia, the picture of progress offered me as a child that is today so laughably "retro." I could say that, in these media-manipulated times, irony proves me smarter than television, more wily than advertising, a thinking individual rather than a mass consumer. I could admit (most truthfully, probably) that I am insecure; I dread the embarrassment of misplaced sincerity, of playing the dupe. For whatever reason, there are enough people like me that we are now said to be living in the Age of Irony, a time when it is hip to place "air quotes" around sweetly earnest ideas like faith and community and the soul and the sacred.

Just when I began placing air quotes around God and the Virgin Mary and the religion my mother wanted me to have, I am not sure. I recall no agonizing crisis of faith. I did not so much lose my Catholicism as casually shrug it off, leaving it there in the pews like a forgotten sweater on a warm California Sunday. This may have begun as early as age ten or eleven, when my mother would fret over whether to let me eat a hamburger on a Friday (then still forbidden) and I'd crack, "Oh, Mom! Nobody's going to hell on the meat rap!" When I put it that way, she could not help but laugh along with me. I know that, barely a teenager, I listened to the catchy new folk hymns meant to evoke sweet earnestness in me, and I was glad for so soft a target for my sarcasm. *Kumbaya my Lord, kumbaya . . . Yeah, right.*

I know that by the time I was attending Catholic high school, mine were the surlier expiations of an Ironic Fundamentalist. One dull afternoon I wandered into the chapel with a few friends (one of whom intended to become a priest) and, with no one else around, I suddenly waggled my middle finger in a fuck you salute to the Christ on the wall. My startled friends searched my face. Was I angry at God or just sure He was dead? Would I pretend such a vulgar atheism just for a laugh? Or was my point that *my* God, a profoundly un-Catholic God, was so impossible to offend that He laughs back?

I left it to them to decide. I didn't bother to explain it even to myself for I had accomplished my only aim, which was to puncture the dullness of that afternoon with my new plaything, my ever-sharpening sense of irony. If I was reckless with my recent discovery, it was only because I was so very pleased with it. For how could you be caught inside a banal moment if you also stood outside of it, always poised to destabilize it? How, for that matter, could you be trapped in a banal self if you hovered above your own life, suspended upon layers and layers of ironic detachment? That, to me, was the saving power of irony. Staring back at my friends in the quiet of the chapel, still flipping off Jesus, I waited for a wave of remorse to well up in the former altar boy. But waves of feeling, when filtered through layers of irony, have a way of flattening into a confusion of small, ignorable ripples.

These many years later I must confess that I lack, still, a sinner's guilt for having shrugged off Catholicism. And yet here I am, sitting next to my mother at Mass, here to sample the Catholicism of Queen of Apostles today, a Catholicism that is attracting the return of many a fallen-away son and daughter of the blue sky suburbs. This Mass, I have been told by my mother, is the most popular one for miles around. The pastor who packs them in is Father James Mifsud—Father Jim as he likes to be called—a bald, compact man who grew up in a big working-class family on the rough side of San Francisco, a priest who lifts weights every morning and who proudly proclaims that his Jesus "is not some wimpy guy with no muscles," a priest who has made himself welcome in the dugout of the Giants baseball team and the locker room of the football 49ers. Father Jim's famous sermons have, in fact, the flavor of a rousing locker-room harangue, peppered as they are with swear words and shouts of encouragement and even the occasional ironic joke.

"My mother taught me: Every time you lie, blue lines come into your forehead." Over a flutter of chuckles, Father Jim's voice swells.

"I said to her, 'Blue lines? What blue lines?'

She said, 'Only your mother can see the blue lines.' So. I grew up learning not to lie . . . *because my mother lied about those blue lines!*"

Grateful laughter pours forth from his audience, including me, as I absorb the latest modernization of a church interior that once seemed so modern. How much brighter and more streamlined Father Jim has made Queen of Apostles by installing racks of track lighting and by painting white over all the knotty pine and cinderblock. The altar has been lowered and moved forward so as to make the Mass (as Father Jim explained when he did it) more democratic, more a celebration of community. The enormous, agonized Christ who once hung on His Cross on the back wall has been spirited away. The bared space is used for projecting words to hymns so that all can sing along accompanied by whatever image of God each prefers to keep in one's head.

"Everybody," thunders Father Jim, "is welcome here."

Everybody is welcome in the community of fragmented opinion that American Catholicism has become. An "attitude problem" as *Newsweek* put it, grips a Church "honeycombed with groups who want more: the democratic election of bishops, optional celibacy for priests, a declaration of rights for dissenting theologians and blessings on monogamous gay marriages." Two thirds of American Catholics say the Church is wrong to oppose birth control, the same percentage thinks women should be priests, and even more are for ordaining married men.

One in ten Catholics told a *New York Times* survey they rarely if ever pray privately. One in eight said that to be a good Catholic one needn't even believe that Jesus was the Son of God. Almost two thirds of American Catholics, and half of regular churchgoers, dare to see the central sacred ritual of Catholicism, the Eucharist, as a symbolic flourish. Apparently, in the communion line of Queen of Apostles Mass, there are many who answer *Amen* to the declaration *Body of Christ,* all the while placing air quotes around their assent.

Father Jim sidesteps all these fault lines by fashioning a community around what his flock shares: a sense of the grip slipping. Father Jim speaks not to the old future I grew up with, but to the latest one, the future that won't let you take a thing like progress for granted, the future that makes you wonder all the time whether you are good enough to hack it. "You are better," goes one of Father Jim's favorite pronouncements, "than you think you are!"

"Never Give Up!" goes another. Outside, draped above the church entrance, a vinyl red on white banner proclaims *Never Give Up!* This has become the public theme of Queen of Apostles parish, visible to anyone driving by. It is printed on signs handed out for free. *Never Give Up!* It is emblazoned on the parish bulletins in every pew. *Never Give Up!* It is urged by Father Jim in his pulpit on the day I am there to hear. *Never Give Up!*

I look around at the sea of people who have come to hear this exhortation at a time when the blue sky dream is being downsized. A surprising number seem to be like me, soft and white and in their thirties, dressed in their casual Sunday best. I hear Father Jim demand that all of us pray together, for each other, loudly and with feeling, the way a community would. Someone stands and says, "For a special friend with a drinking problem." *Lord, hear our prayer,* we all answer. Someone says, "For my sister who's having legal problems." *Lord, hear our prayer.* "For Dad's speedy recovery from a stroke." *Lord, hear our prayer.* "That those who are unemployed may seek work with faith, hope, and patience." *Lord, hear our prayer.*

Later, as I listen to my mother sing her hymns so happily and charmingly off key, as the Mass nears an end, I realize that a moment I have been vaguely dreading is at hand. It comes when my mother stands to join the line receiving Holy Communion and I remain seated in the pew, drawing my legs in to let her pass by.

Regardless of what *New York Times* surveys say, the Church teaches that through the priest God literally transforms bread and wine into the body and blood of Christ, so that to take Communion is to participate in a fresh, recurring miracle. My mother is one of those Catholics who still believe in this miracle. I, of course, cannot. And so, Christmases, Easters, every Holy Communion brings the same dilemma. If I were to follow my mother into line, say *Amen* to the priest's *Body of Christ,* swallow the wafer, and squeeze my eyes shut, would I be telling my mother what she hopes is true, that I am, at least, wishing to believe? Or would I simply be dishonoring her own authentic belief? Each time, I wonder this. Always I have chosen to remain behind and sit and leaf through a hymnal with my head down, like a criminal suspect not wanting to be noticed.

This Sunday morning at Queen of Apostles is no different. Except this time, as my mother makes her way up the aisle toward the body and blood of Christ, I do not find the usual solace in detached irony. I begin to imagine my ironic sensibility as a gremlin who sits on my shoulder and whispers dirty things into my ear. I see my mother take Communion, have her miracle, and I think: What if the essence of an ironic life, the striving to live both inside the moment and outside of it, is merely to live twice removed? What if, instead, I had managed to hold fast to the Catholicism that everyone around me in church seems to have kept, somehow, on their own terms? How must it be, to stare at that sparkling new baptismal font and not see a Jacuzzi?

"I don't think Queen of Apostles parish has changed much in thirty years," Father Jim told me a few days later. We chatted in the study of the ranch-style rectory that sits across the street from church and school. "I think this is still a white middle-class neighborhood."

Father Jim was well-acquainted with the history of Queen of Apostles; he celebrated Masses here in 1965 as a newly ordained priest. Even during the thirteen years he spent as a missionary in Seoul, South Korea, Father Jim maintained his ties to Queen of Apostles, returning every summer with slide shows and requests for donations. When he became pastor in 1989, he was the old friend returned home. But when Father Jim said that Queen of Apostles has remained white and middle class and therefore not much changed, he was of course wrong, because to be white and middle class in this neighborhood today means something very different than when I was growing up.

Once, this was a frontier parish on the edge of an expansive idea called California. That was when the four-bedroom houses cost twenty-five thousand dollars, no money down, and Lockheed was hiring for life. But these days those same homes sell for half a million dollars, and realtors say the neighborhood is "highly desirable" for its peaceful remove, and the good public high school that, as everyone has been told, can make or break a child's future in the new economy. These days Lockheed is laying off and no Silicon Valley firm pretends to hire for life; just until the next "right sizing." These days, around the perimeter of the parish, at the freeway on-ramps, homeless people hold signs saying "Will Work for Food." The white middle-class people of Queen of Apostles are the Californians who frown as they watch the rest of California on the six o'clock news, the California of too many Crips and Bloods, too many drop-outs, too many have nots, too many "aliens." A once expansive idea is now imagined to be collapsing from the edges in, and Queen of Apostles has changed from a frontier parish to an enclave of worried, tenuous affluence—without room, even, for the children who grew up here. "I see a lot of kids getting married and wanting to move out of California," Father Jim told me when I asked him what troubles he saw in his parish. "It's almost impossible to finance a home.

"The other big negative, I find out from my parishioners, is the difficulty of maintaining a position in highly competitive companies and society. Extremely difficult. A lot of cutthroat." Father Jim shook his head. "That does something to you. It drives them to an awareness of their faith. Their religion and their beliefs are quickly tested by the realities of life and if they don't have that kind of a background, they despair quite easily. To be constantly competitive in order to stay financially well-off can cause one to feel *very* lonely and despairing.

"I think the challenge for me," said the pastor, "is to get people to believe in themselves. They are not sinners, they are God's children. They want to hear something *positive*. They want to hear something *human*. Stand *up* for yourself. The jackass boss, the troubled marriage, the autistic child, the friend dying—they need strength for those things. That's what they're looking for, I think. They want to know God loves them. That they're okay. That they're better than some people say they are. And once they begin to believe in themselves, they will open themselves to others. So I work on telling people they are good, they can be better, and there are other people who need them."

His delivery grew more staccato as he warmed to his theme. "I tell 'em: 'Instead of sitting on your *rear ends* and watching the problems of society unravel on the six o'clock news, what are you doing to become involved in some way? And the parish must be the hope that provides people with opportunities. *That's* why we have the employment program here.' "

In a refurbished church storage room a retired realtor and other parish volunteers clip classifieds, make phone searches, and otherwise help jobless people hook up with employers. When

Father Jim started the Ascent program, he intended it mostly for ex-cons he'd meet in his jail visits, though lately a lot of out-of-work aerospace engineers had been showing up at the door. As for the homeless people at the freeway on-ramps, Father Jim had a word for them. "Crooks, most of them."

"Crooks?" I said, startled by the word, so damning in the mouth of a priest.

"God, yeah. *Crooks*. Every one within the radius of this parish I've bumped into, I've given them a card. I say, 'You're fifteen minutes away from getting a job. You don't have to do this anymore. Come. I'll take you there right now.' They say, 'Well, I'll see you later, Father.' No one has ever come. Why should they? If they can make a hundred bucks a day from some suckers who give money to them, why should they? They're not gonna make a hundred bucks a day in any kind of a job." For this form of Silicon Valley entrepreneurship, Father Jim has no sympathy at all. "And then they give you the Vietnam crap! And they have babies sitting next to them! Most people that way are dishonest. They've got arms and legs. They can walk. When people call here for help, I make sure they get help. God helps us to help ourselves. Well, that's what I want to do. Help people to help themselves.

"I'll say to someone, 'You've been coming in here the last three months for food. Have you got a job?' "

Father Jim mimicked the mumble of the beggar. " 'Well, uh, no.'

"Then get a job! And don't come back here until you've got a job!"

"There are jobs to get?" I asked.

"Sure! We put fifty to sixty people to work a week."

Tough love for the panhandlers, pep talks for the beleaguered executives, and a job program that maps and fills the cracks of the new economy, promising a place for everyone. That is Father Jim's pastoral recipe. The anxious middle class of Queen of Apostles finds strength in the assurances of their blue-collar priest, his message that when things go wrong, nobody's perfect so just keep trying. The flock is appreciative of a priest who wants them to believe, above all, in themselves.

I told Father Jim that despair, the deadliest of sins, is the one I found impossible to comprehend as a child in sunny, booming California, no matter how many times the nuns tried to explain the concept to me. Do people who live here now so need to hear, over and over, his *Never Give Up!* refrain?

"The people who commit suicide," he reminded me, "are not those less well off."

After my conversation with Father Jim, I strolled over to the little cinderblock complex where I spent grades two through eight. In the 1960s, Queen of Apostles school had the air of a mission outpost, so sparse were the facilities, so poorly paid were the lay teachers (many of them unaccredited). We endured such shortcomings because, within those cinderblock walls, Catholic doctrine was so concentrated as to be sure to permeate our souls. Why else would we not attend one of the shiny, sprawling, new public elementary schools?

Marianna Willis, who taught fourth grade when I was in eighth, had stayed all these years at Queen of Apostles and become principal. I found her in her office, her face as kindly as I remembered, her Kentucky accent still softening the edges of her words. She said, "When you were here, remember how the racks were full of bicycles every morning? Now there are maybe three. This used to be a neighborhood school. Now it really isn't."

The students live as much as an hour away in towns like Gilroy and Morgan Hill, where tract homes are still being built on garlic fields. The parents commute into Silicon Valley and drop their kids off before work. That is why Queen of Apostles now operates a day care from 6:30 in the morning to 6:00 in the evening. I smiled to remember my mother riding her bicycle over to school with a basket full of hot dogs for her children's lunches, my mother with another mother placing a pink frosted cupcake on every desk for a Valentine's Day party, my mother and all the other mothers taking their turns as volunteer supervisors of dodge ball and jump rope games at the lunch recess. Marianna Willis recalled the same era, when Queen of Apostles children came to school on Mondays with stories of picnics and excursions taken with Mom and Dad. A few Mondays back, she asked a girl why she was excited and this is what the principal heard: "Oh, my mom got to sit down and eat dinner with us."

Monthly tuition, once $15 for me, is now $256. There is financial aid for only ten families and "a lot of our people have lost jobs, or are single mothers," said Marianna Willis. At that, applicants are turned away. Enrollment is highest ever and full at 301, a number that includes "a handful of blacks, some Hispanics, quite a few Asians."

Marianna Willis told me what Queen of Apostles parents today were wanting for their money. "Academics is a big thing," the principal explained. "Quite often a C in public school translates to a D or F from us." Low-performing applicants were therefore screened out by an entrance exam. Another selling point was peace of mind. "When you hear about the guns and knives in the public school, you can't help but think, 'I don't want my child there.' " The worst such scare Marianna Willis had faced was the confiscation of a rubber band gun. And unlike public schools, Queen of Apostles need not keep children with disabilities or behavior problems. "We don't have counselors. Would it be fair to take one child who's going to take half the teacher's time? We have to be able to serve the community we have."

That community of carefully culled children is not necessarily a Catholic one. Some students at Queen of Apostles are, for example, Buddhist. Still, another reason parents of various faiths send their children here is the Catholic ethos, not just the prayers and Masses, but the good deeds demanded of every pupil. Marianna Willis described for me the regimen of beneficence: the kindergartners baking cookies for the convalescent home, the first graders writing notes to the sick, the fifth graders visiting disabled children, the other grades preparing and delivering food for the hungry. None of this went on when I attended. Our outings were to see a soft drink bottling plant or the mummies under glass in the local Egyptian museum. Once, I think, my class adopted a poor child in some lost land like Bangladesh.

"We teach the children to live a Christian life, to make religion a part of their daily experience. We teach care and concern for others. We try to teach our children they are part of a community," said Marianna Willis. What the gremlin of irony whispered to me is that Queen of Apostles school was from the beginning a rather ruthless protectorate, a nun-run Utopia dependent upon a military contracting economy that allowed mothers to stay at home and fathers to keep jobs and children to grow up cocooned from the implications of our great luck. Today, with no nuns but tough screening tests and pricey tuition, Queen of Apostles school is at least more frank in its ruthlessness, its children, the fortunate 301, reminded of their brightly diverging prospects whenever they deliver lunch to the ever-full Emergency Family Shelter downtown.

Later that morning I sat in the back of the seventh-grade class sampling, along with the budding teenagers, another example of what Queen of Apostles had to sell that public schools did not. Father Jim was roaming the front of the room, splintering chalk against the blackboard whenever he turned to write a word like "RESPECT."

"Here's what I want, you little farts. (Nervous laughter.) I want some respect. If you brought a friend from a public school and you were walking around here and you threw a piece of paper on the ground, I wouldn't call you a jackass in front of your friend. Because it would embarrass you. I'd call you over and say, 'Hey, don't you have respect for this property? Good, then go pick up that paper.' You'd give that guy a bad impression and you'd give me a bad impression, but I would never do that in front of him, make you look bad. Because he's your friend. That's called *respect*. That's called *respect*. I'm talkin' about all different kinds of respect. I'm talkin' about respect for authority. People who love you, who are older than you and work for you. I'm talkin' about your parents. I'm also talkin' about respecting yourself. Get this in your *heads* when you're *young*. So you flunk a test! No test is worth cheating for. Big *deal*. You're not going to the electric chair. No test is worth cheating for. Look at the people I visit in San Quentin. Here's a guy who gets mad one time because he's drunk, comes back and shoots a kid in a gas station and kills him. He goes to jail for the rest of his life for one stupid ass thing. But that didn't happen because of just one time, it happened because he *stole* things when he was little, he *lied* when he got older, he *cheated*, then he got into the *drug* business. That's what I'm trying to get into your thick skulls now. Hey! When I'm talking, I want your eyes on me! No place else. Joshua! Are you deaf or are you just thick? I'm not here talking for my health. I'm interested in you because I love you and because these are the things that will keep you a decent human being and you'll learn them *now!* Otherwise why the hell are you paying 2,500 bucks to go to school here when you can go to any other school that has good teachers and a good curriculum and much better facilities than we do?"

At this point Joshua, and the rest of us, were surely giving our full attention to Father Jim, who had known convicts in hell, who could summon a vision of a descent from the middle class into hell, a vision of damnation more tangible than any the church of my youth could have made me believe.

"Did you know that Mary appeared in Medjugorje just before the war in Bosnia?" my mother asks me. She reminds me that Medjugorje is a village sixty miles from Sarajevo, the city where evil currently resides most vividly on the television news. "She tends to appear to people who are really discouraged. And she always appears to poor people. She appeared in Ireland before the potato famine. Did you know that Immaculate Mary is the patron saint of the United States?"

My badly suppressed grin as she tells me this does not seem

to threaten my mother's belief in the slightest. She is hardly an irrational person. A trained scientist who returned to work in a medical laboratory after her last child reached high school, she reads the newspaper closely and sends me insightful clippings about the latest political and economic shifts in California, America, the world. At the same time, she is unshakably sure that Mary has "a plan for world peace," a plan the Blessed Mother is in the process of revealing through mystical appearances.

This feminine face of heaven, this Mary who shimmers into view now and then to reassure the world of her concern, is central to the Catholicism my mother wanted to give her children. Our Lady of Fatima, Our Lady of Guadalupe, Our Lady of Lourdes, Our Lady of the Potato Famine and the Bosnia End to a New World Order, all of them Ladies quiet voiced and lovely, without wrath or vengeance. When the mysterious Lady appears, it is to tell us she has a plan for the world that is beyond our ken, a plan that will gradually, miraculously make itself apparent if enough of us pray hard enough.

This is the Catholicism, of course, that was most vulnerable to my ironic obliteration, the notion of a supernatural otherworld populated with saints, angels, three Gods in One, and a Virgin Mother mysteriously worrying the world toward a better future. This is a Catholicism that Father Jim likes to downplay in his message to Queen of Apostles today.

"The mystery experience," Father Jim told me, "is not all it's cracked up to be. Most people can't handle the mystery experience and they tend to go off the deep end, become very fanatical in their religion. Mystery," he said, bringing things back to the personal and pragmatic as he likes to do, "doesn't mean something that's unknown. Mystery is a deep appreciation of the goodness of the other person, whether it be God or someone else. I think that is what should be stressed, rather than the fact of the unknown. Because unknown things tend to frighten us."

Still, as much as my mother enjoys the down-to-earth preaching of Father Jim, she also elects sometimes to leave this world, to inhabit, through prayer and belief, the shadowy, mystical world that the Catholicism of her girlhood made available to her, a world she tried to give me and I laughed away.

Hail Mary, Mother of God. My mother prays to the Queen of Peace who appears, now and again, to six children in the mountain village of Medjugorje. "I have come to tell the world that God exists," Mary is said to declare (in the pamphlet my mother shows me.) "He is the fullness of life, and to enjoy this fullness and obtain peace, you must return to God."

Tony, Tony, listen, listen.
Hurry, hurry, something's missin'.

My mother prays, still, to a St. Anthony who has been granted the miraculous power to turn up what has been lost.

In the hope of feeling cleaner in her presence, I make an appointment to tell my mother I am a pagan. I ask her to lunch, driving her to a pleasant bistro of her choosing, disclosing, before the dessert arrives, the uninvited. I tell her I respect her faith in God, but that I have none. She is quieted, peers back at me with calm concern. Then, in her measured, elliptical way, my mother tells me, in so many words, that she prefers to have faith that I will have faith.

I ask her, "Do you feel embarrassed when you pray for me, asking God to favor someone who has betrayed your religion?"

"No!" she says, suddenly brightening, smiling. "God loves you. God loves all of you." She is grouping me with my brother and two sisters and father, all of us who have said, in varying degrees, no to my mother's Catholicism. "He sees the good in you."

Her belief in a God who is willing to wait me out reminds me of the many times I've told her over the phone, "Don't waste your prayers on me, Mom, I'm doing fine." In our sun-kissed way, we children have always kidded her whenever she has voiced her worries, demanding that she be her usual good fun. "It's not that I don't believe in you," my mother would usually clarify. "It's just that so many things can happen to anyone at any time." This, I now see, is an apt summation of an awareness in my mother, a tragic awareness that, like her belief in miracles, she has retained despite the smooth brightness of the suburbs.

I begin to realize, as the dessert comes and we eat it shyly, that my mother has somehow resisted becoming a creature of California's famed optimism—the optimism that now is curdling into a bitter tantrum of thwarted expectations, the angry people of the suburbs who say their future was ruined by welfare mothers and immigrants and liberal politicians. I begin to see that my mother's attraction to a cheerful and casual Catholicism did not cause her to forget the catechism of the Depression she learned as a girl in Rock Island's dark old churches. My mother has always understood that fate is lean and mean, that anyone's future could be drastically downsized at any time. Yes, hers was a God who unreservedly endorsed progress, personal and national, and the good suburban life that went with it. But never would He *guarantee* such progress, and only a people of hubris would think they could guarantee such progress for themselves, by themselves. All of my mother's religious folkways in the midst of aerospace suburbia—her statuettes of Mother Mary on windowsills, her cataloged powers of the patron saints, the Styrofoam Advent wreaths she helped us make every Christmas—were like candles lit against a darkness that, should it come, would not surprise her, would not extinguish her faith.

And so, quite the opposite of a shallow optimist, my mother has been, all along and still, a person of Christian hope. Such hope begins by acknowledging the forever frail helplessness of every soul and, in that acceptance, finds all the reason to believe in a merciful, all-powerful God. A hope like that humbles a person, makes a person generous in the face of human stupidity, fashions faith from anything, nothing. Such hope is at the heart of Christianity's great paradox, the most difficult of ironies, the acknowledgment that we are saved by knowing we are lost. Having shrugged off my Catholicism, having flipped off Jesus Christ, I am unable to fathom my mother's hope. I will have to content myself with godless ironies, easier ironies, and that, I suspect, is my great price to pay for having said no to my mother's Catholicism.

"As long as we're near downtown," my mother says, "let's drop into Sacred Heart." She means the place where she has volunteered for the past few years, Sacred Heart Community Center, a whitewashed storefront on a particularly desolate corner of old San Jose, fifteen minutes on the freeway and a world away from Queen of Apostles. When we arrive there a tanned, vivacious young woman named Elisa gives my mother a big hello and whirls me on a tour of racks of secondhand clothes, stacks of donated food, the closet-sized free health clinic, the computer training room, the daycare. "Our clients are immigrants, the working poor and urban migrants who saw one too many sitcoms about California and came out here with no support system," she explains. "We fed sixty-six thousand individuals in the course of last year."

Elisa was born a little over a decade after me and grew up in a house like mine not far from mine. After college, her boyfriend took a job opening up a new Gap store in Moscow while she headed for L.A. and landed work in rock 'n' roll public relations promoting Depeche Mode and U2.

"But I felt a tug," she says. "After a year I came back and wanted to do work in service." Elisa tells me about the founder of Sacred Heart Community Center, a woman named Louise Benson who, at age sixty-one, began gathering and giving away clothes out of her tract home garage until complaining neighbors made her move the operation. "Louise couldn't sleep at night knowing people were going hungry. And nobody worked harder. She always said, 'Minister with dignity, compassion and respect.' " When Louise Benson died at age eighty-one, eight years ago, she left, says Elisa, "so many real stories of grace about her."

"Grace" is a thing Elisa knows and wants. It is why she

comes here to work. "It's nice," she says, "to see the action and the grace around you." "Grace" is one of those words Elisa and my mother say often and with ease, a word I would feel comfortable saying only after placing around it air quotes. Later in the afternoon, my mother invites me to drop in on the first day of the Survival English class she helps teach at Sacred Heart. I listen to the new students introduce themselves for the first time in a new language.

My name is Ramon Luis! My name is Irena!

Head teacher Sister Elizabeth Avalos, in a flowered blouse and purple skirt, is making everyone shout the words out with silly gusto, as only children usually are made to do.

My name is Fidelia! My name is Hector!

I count more than a hundred people jammed into the little room, and I get a sense of why my mother makes her pilgrimages here. She has come to be with people who find hope in being in California, who might even believe what we believed when we first came to the new subdivisions of Queen of Apostles parish: that in California the future could be annexed and developed and built out to make room for everyone.

My name is Virgilio! My name is Juanita!

I sit on a box in the corner, the interloper, vaguely ashamed that I should find their expressions so foreign. I would not want people to see me with that face, so open, naked. I think to myself, here is a place, an experience, immune to easy irony. And I realize that this has been my mother's secret resource, the reason she has always been able to laugh patiently at my blasphemies and to be a modest mystic amidst a sterile subdivision. She understands how to seek out and fit within moments immune to sarcasm. She is at home within these moments, and maybe that is to know grace.

Tony, Tony. Listen, listen.

No prayer of faith, in my mother's mouth, could sound corny.

Hurry, hurry. Something's missin'.

In mine they all do, and I expect they always will.

On the last night of my visit home I look up an old college buddy who has become a successful corporate designer for Silicon Valley. Like some of my other Ironic Fundamentalist friends, he has decorated his house with ornate Catholic icons, saints and crucifixes and shrines handpainted by peasants. The son of an Italian and a Colombian, he grew up living all over the world, Brazil, Mexico, Cairo, Rome. Having been smothered in a baroque religion, he now finds its images beautiful in a playfully agnostic sense, and as usual, he was ahead of the style curve, anticipating by many years the brass and crystal crosses that Donna Karan now sells as $395 jewelry accessories.

I arrive at the door expecting a lot of vodka and laughs, but the news is terrible. My friend's mother is dying quickly of just discovered cancer. He sits on his porch, eyes away, weeping in the dark. I don't think he wants my arms around him, but what can be said?

I say, "Does she believe in God, in heaven?" He turns toward me, his face a wet crumple, and we two friends who are faithless look at each other and are grateful for his answer. "Yes."

Nearly a year further on in the Age of Irony, I pull from my mailbox one of my mother's packets of clippings. This long after my confession of faithlessness, her letters have continued to arrive as newsy and funny as ever, nothing, seemingly, having changed between us. These many months later, I have done not a thing to reward her faith that I will have faith. I have not sought out a Catholic Mass here in Vancouver. I have not wanted Father Jim's macho common sense, his Mass as motivational seminar.

Nor have I wanted the old-fashioned Catholic Mass I'm told is still celebrated in these parts, the mystical supplications, the whispers inviting of the feminine grace of a Virgin Mary. I am wanting to stay out of churches altogether for the reason that I find it a trespass to pray with strangers, to pass myself off as their community, all the while silently editing our prayers in my head until all is metaphor bracketed by air quotes.

Nor have I, when alone, attempted to send a single prayer to heaven, even though I know my mother would want me to try. The part of me hovering above, detached, would be winking at my own halfhearted attempt and I couldn't bear that. The conclusion I have come to is this: Once the ironic instinct has scoured away the sweet earnestness of religious faith, the only recourse is to acknowledge the loss with a more rueful sense of irony.

For my mother I feel the same envy I felt that day in church as I sat and watched her step forward and receive Communion, her weekly miracle. She has kept what she brought with her to the suburbs before she ever set foot in Queen of Apostles church. She has kept her mystical imagination, her own way of defeating the banal, of living at once inside the moment and outside of it.

Her mystical imagination allows her to believe, when the latest news of a dream unraveling comes from Sarajevo or Silicon Valley, that this means the Virgin Mother is all the more likely to reveal her restorative plan soon. I refuse to scoff at that belief for the simple fact that I love my mother too much to demean the faith that is at the center of her being. And if I would not demean my mother's belief, who am I to scoff at the belief of anyone else who might also happen to know grace? So, after all her many prayers for me, you could say that my mother has given me, if not the Catholicism she wanted me to have, at least this. She has made me want an ironic sensibility that is more generous and patient, more content to trade hope for optimism, more like her.

I open the packet my mother has sent me, finding a short article from the *San Jose Mercury News* written by a staff writer named Tracie Cone. The writer is waxing amused at Our Lady of

Peace, the 32-foot, 7,200-pound stainless steel statue of the Blessed Mother that stands overlooking a busy stretch of freeway in Silicon Valley.

The writer finds very funny the thought of a Mary "blessing perpetual gridlock and the souls of hot-footed drivers."

The writer has found a believer, a woman who answers phones at Our Lady of Peace Church, to say of the sculpted Virgin, "In a lot of ways she represents Silicon Valley. Stainless steel is so modern and this valley is so state-of-the-art."

The writer has discovered something delicious. "At Our Lady of Peace, souls seeking solace may have their prayer wishes recorded on microfilm and perpetually stored inside Mary's giant heart, much the way tax records are stored at some county office buildings. . . . So far, more than 100,000 prayer requests have been loaded, with room for millions more."

The writer has made a bit of room for the perspective of Sister Mary Jean, who is secretary of the shrine. "This is a time of great tribulation. People release sorrow just by putting it on paper. They know God's mother understands."

The writer has summed up in the voice I instinctively recognize and warm to, the voice of the gremlin who whispers in my ear and will not be brushed away. "The church folks say Mary will receive your message," ends the column, "even if heaven is a place without microfiche machines."

Attached to the article is a note in that familiar, small script of my mother's. "I thought of you," she has written.

ELEVEN

ORGANIZATION MAN RETIRES

My father and I are in the sky again. We have not been here together since a day twenty years before when he had placed me with him under the bubble canopy of a sailplane and, to show me what could be done with no power other than a thermal updraft, had sent the two of us somersaulting over the soft green ridges east of Silicon Valley. Shortly after that day, my father stopped flying. He devoted more time to the family, or else, whenever he wanted to be alone, he spent hours in the garage repairing television sets and stereos for friends and neighbors. Those fix-it jobs held a double attraction, he liked to say. They were problems he could solve, and they afforded a chance to make someone happy. But such satisfaction, I sense, could not match his enjoyment now as he pokes the nose of the Cessna through a stratum of cloud above the Santa Cruz range. "As I used to tell the other guys in the squadron," I hear in my headset, "God must have intended man to fly. Why else would He have made the tops of clouds so much prettier than the bottoms?" I smile in surprise at my father's uncharacteristic sensualism, his even rarer talk of God. He is smiling, too. "They all laughed at my little attempts at philosophy."

It was shortly after his early retirement from Lockheed Missiles and Space Company that my father took up flying again. He was only sixty years old, but after thirty-two years with the company his salary and benefits had grown so that the downsizers of Lockheed reckoned it would be cheaper to pay him to quit. They offered him and several thousand others a deal sweet enough that to say no meant you either needed the extra bit of pay that staying on would provide, or else you needed to prove you were still needed by Lockheed and the working world. My father did not place himself in either category. Besides, he guessed (accurately) that the deal would soon evaporate, and people like him would be shoved out the door without the cushy incentives.

So my father was lucky at the end, as he had been lucky (he was saying often, now that he was retired) his entire life. Lately, he would lay it all to "flukes of chance": He was a teenaged "airhead" handed a Navy scholarship; the Navy "happened" to see in him a jet pilot; "by plain dumb luck" he joined aerospace at the dawn of its Cold War heyday; he had "the great good fortune" to meet my mother; they arrived in an unspoiled California where "the great joys of his life," we four children, "flourished"; certain investments "worked out"; when his employer no longer wanted him, the news came at the "best possible moment"; and now, upon entering retirement "with health and finances intact," he has found "the perfect thing" to fill his time, a membership in the Seagulls, a group of men who share ownership of a Beechcraft Bonanza and this Cessna 172 he is piloting today.

"Well, nothing pressing on *my* calendar," he says as we drop into more undulating whiteness, emerging to find the Monterey Bay silver-blue and calm before us. "How about we land in Watsonville for a Mexican lunch, do a little whale watching on the way home? Sound like a plan?"

Among the unlucky was Gary Kolegraff. Hired out of college by Lockheed to be a cost analyst on a satellite program, he was laid off ten years later at the end of 1993. He had never married. At thirty-five years of age, he was renting, together with a government geologist and another Lockheed worker facing layoff, a tract home not ten minutes from that of my parents. Sure, fine, he said when I called a few months into his joblessness. Come on over, and I'll tell you what it's like for me. His chipperness at the door, his face unlined by worry, made me worry for him. There was a shell-shocked artlessness about Gary Kolegraff, the smile of one who hasn't quite comprehended an insult thrown his way.

"When they let you go, it's amazing the cycle you go through. Initial euphoria: 'I'm free!' You're out in the sun playing. But then, I didn't realize what I'd left until it was gone. The paycheck was gone."

This house he had lived in for five years, a prototypical rancher of the 1960s, was never meant to be inhabited by single men with futures clouding over. The bright functionality of the original vision was masked by old green shag on the floors, Formula One posters on walls, fried bacon smell in the air. The living room couch faced a pile of stereo gear tangled in cables and extension cords. In the backyard, neglect had made everything brown. In the kitchen, a magnet in the shape of Lockheed's supersonic SR 71 Blackbird held a Mark McGuire baseball card to the refrigerator door.

"I saw myself as buying a house, settling down, getting married. I guess I saw myself a lot like my dad."

I had found Gary Kolegraff on a list of the unlucky (ex-convicts and aerospace workers, mostly) kept in a three-ring binder at the job search center at Queen of Apostles church. When I saw the Kolegraff name I recognized it. Though I didn't know Gary, who was a few years behind me in Catholic high school, I knew the Kolegraffs were fellow tribe members in the parish. I was not surprised when Gary told me his father had been at Westinghouse for thirty years, a draftsman on nuclear missile launcher designs and other military projects.

"I'm the kind of person who really likes security, routine. Put in your eight-hour shift and come home and have fun. But now it looks like that's changed. I'm looking for any job I can find."

A mere three months' worth of unemployment checks remained, and still Gary Kolegraff was spending a lot of time on the grounds of Lockheed, still there with unlucky others at the Career Transition Center, a place to print out résumés and make calls and wait for the calls to be returned.

"They try to make it happy. They have coffee and everything, but it's not a happy place. Looking around at everyone else just reminds you of all the competition for jobs. It's weird, though. Just showing up at the career center, sitting in a module with a desk and a phone, you feel like you're still going to work."

He was hearing that defense workers had the reputation of being out of touch and even lazy, that Lockheed was ridiculed as a training ground in today's job market. He was imagining the word "Lockheed" read and rejected by those electronic eyes the bigger companies use, the machines that scan résumés before any human being ever sees one. He had lost count of the hundreds of résumés mailed, calls made. All had yielded no more than a couple of dead-end interviews, and chipperness was becoming ever more exhausting to maintain.

"My voice has changed from the stress, a tightening in the voice box. That is showing up in my interviews. It's almost like a spiral situation. I know it will have to change, something will come up. But it's something I have to deal with."

Gary Kolegraff was beginning to think, "I have to almost redefine myself." He was hoping he could learn to be what the people at the Career Transition Center said he needed to be, a "networker." He was hoping he would not have to go about redefining himself while living, again, in the house where he spent his childhood.

"If I can't find a job, I'm considering moving home. That's a reality I have to consider."

Gary wanted my phone number. "If I'm feeling blue or something, can I give you a call?" he asked, and I said sure, fine.

What a frightening specter for a blue sky mother and father: the boy at thirty-five hanging up his gray slacks and striped ties in the bedroom where they used to tuck him in at night. Had the mother and father been like Gary's, had they been gratified to see their child begin a career in his father's image, what would they now say to the "boomerang" returned to their breakfast table? What could they say, except that a parent's duty is to prepare the child to make one's way in the world of work, that they had tried in good faith, but had been mistaken in their teachings?

The lessons, very different, my father taught me about the world of work he conveyed in stories told in the afternoon heat after a lawn mowing, or when we two were alone in the car, or times, after dinner, when he and I lingered at the table. Any of the luck he speaks about today was not to be found in those tales, which I knew were crafted to warn me *away* from a Lockheed career or anything like one.

If, for example, I were to ask him about his start in aerospace, he would tell me about the J79 jet engine of General Electric, and how one of his first assignments was to write a report every morning about the testing progress of the J79, and how there developed a problem with the fuel control at idle speed, and how a device called the Idle Instability Fix was designed and tested to solve the problem. He would say that the Idle Instability Fix had not worked, and so he had tried to win the smiles of his new peers with a note of comradely humor, adding to his morning J79 report the observation that the failed Idle Instability Fix was "Some fix!" He would say he sent out the stack of morning reports only to find, not long after, the angry face of a superior hovering over his desk, informing him that he would now rewrite the report with his two-word editorial gone from the page, that all the originals would be recalled and destroyed, that he was never, ever again to stray from the facts, only the facts, in his morning reports.

He would place his dismayed self, then, in my mind's eye. "We all were at plastic topped, gray steel desks, all of them identical, lined up in columns, mine just a desk in a matrix of desks. If you were busy, it was known. If you were not, it was known. Your escape was to go to the john or the candy machine. We all wore white shirt and suit and tie; we wore our suit jackets all day because we were *professionals*. No self-motivated bright and curious fellow or woman would survive that way today. You had to be," he would say to me, "sort of conditioned."

This story contains most of the cautionary elements of his others, the theme of a man's will bent to the needs of the organization by his superiors' tactics of punishment and reward, the subtheme of the man lacking the awareness (early on) or the résumé (later on) to opt out. "There were strong messages there for me I probably should have heeded" is how my father ends the story of the J79 morning report.

My father began telling such stories to me when I was in high school and pondering how I might make my way in the world of work. He tells me new ones, now and again, even now. I like to think that I have taken his lessons to heart and patterned my life accordingly. I have never been in any job more than a few years. In the one large organization that briefly employed me, the Hearst Corporation, I behaved with a brittle suspicion of higher management. Otherwise I have joined modest enterprises, several of them nonprofit, which offered low pay and little security in exchange for a slack leash on my time and energy. I am today a freelancer in an economy that is said to have less and less need for anyone who is not a freelancer. My siblings have, as well, all in their own ways, reflected my father's lessons: my sister who practiced physical therapy in small clinics here and there before taking time out to be at home with her two young sons; my brother, who is married with two daughters, and is a self-employed eye surgeon after buying into a private practice; my younger sister, who, having earned a master's degree in Spanish, teaches various languages on contract to universities in Spain, France, and Southern California, deciding which international boundary next to cross on the basis of what might prove the most fun and lucrative.

All of us are variations on who you might be if you had "flourished" as a boy or girl in blue sky suburbia but had been given no encouragement to replicate life within the tribe. None of us was ever given cause to imagine a desk for ourself within the windowless walls of Lockheed. None of us today presents the specter of a boomerang child. Gary Kolegraff is the organization son my father and mother never had.

The phone rang and it was Gary, calling my home in Vancouver two months after I'd seen him at his place. He had been feeling a little blue, he said in that eerily chipper voice of his. He wanted to tell me about what had happened at the dinner table of his parents.

"I pretty much broke down. There was like an emotional outpouring of mine. The father of a friend of mine died last Friday and I went to the funeral. I was there and it suddenly hit me. That could be one of my folks. And being laid off, what would I do without them? I've been eating there a lot, pretty much every night now, just me and my mom and dad. And I pretty much broke down and told them how it felt."

Gary had been into the Career Transition Center three times a week since we had talked. The staff continued to offer him coffee and tips on "self-marketing." He was applying for any job that paid ten dollars an hour. He was considering redefining himself as an airline clerk because they used computers. He had quit a data entry class after three days because he was typing when he should be looking for work. Among the ranks of the unlucky at the Career Transition Center, he was noticing a thinning out, a giving up. He was hearing that Lockheed had just laid off another 2,000 workers in Georgia and that morale among those left at the Sunnyvale complex was very bad.

"A lot of backstabbing is going on. A lot of people are working eighty, ninety hours a week from fear they're going to be the next one."

He was spending his time reading books about the stock market in the library of Santa Clara University, though he was not a student there; waiting by the phone at Lockheed, though he was not an employee there; sitting alone in the quiet of his rented tract home with an "incredible feeling of disconnect. You're floating around in space. You've got an oxygen tank, but nothing else." This is what he had been trying to explain to his parents the night he broke down.

"It was almost like a movie. I realized that I had never told my dad that I loved him. I had never hugged my dad. I asked him if I could give him a hug. And I did. It was real moving."

After he left, his father broke down, too. His mother told him so later. The next time he was at the dinner table of his parents, they told him not to worry.

"They said, 'We're not going to tell anyone.' I was afraid my brothers would know about it and think I was crazy. But now I think I shouldn't be ashamed. It's amazing, but I felt better afterward."

He told me this as if emotional catharsis was a phenomenon previously unknown to him.

My father's hugs are such crushings of pleasure that he seems to want to close every seam of light between, to merge two into a single lump of whiskery warmth. When a hug is finished and his face comes back into view, his beaming affection is something his sons and daughters have come to laughingly chide him for, his "dopey grin, *big!*" He has been a hands-on father from the time we were babies, a father who liked a child on his shoulders or on his lap or carried under his arm like a football on the way to the bathtub. All our lives, he was a father more than generous with hugs, except, of course, when he was lost to those bouts of petulance, when those down-pulled eyebrows and tightened lips formed that map of danger we all recognized.

There was the time during my adolescence when those bad moods threatened to push me beyond reach of my father's hugs. His tantrums had become too common and heedless of effect upon me, upon my brother and sisters and mother. Plainly we did not deserve the vehemence of his outbursts, so randomly did they come, so puny were the provocations. Too matter-of-factly did he expect our indulgence afterward, as if a father's authority naturally extended to bullying. Shouting back seemed only to redouble his attack. I was tempted to leave the bully to himself, for I had been told that bullies were best ignored.

What happened instead, however, is that all of us, my brother and sisters and mother and I, entered into a conspiracy of joshing humor against my father, a conspiracy he completed by gradually joining in. We learned to mimic his tics of irritability, the way he poked his glasses up the bridge of his nose, the way he rammed the heel of his hand against his forehead just before exploding. We repeated back to him his own voice muttering, "What now, damnit? I've got a blue million things on my mind!" We invented new names for him, "Daddyman" (a lampoon of Waltons-speak on TV) and "Tuttle" (after the super repair man in the movie *Brazil*). The names, like the jokes, were methods of diminution, a thousand tickles eroding his dour self-seriousness. As we wandered into this strategy, I believe we were encouraged along, at some level, by my father. The best in him always had been able to recognize the absurdity in a situation, even a bad situation, and play it for a laugh. That part of him seemed to welcome our teasing now, as if he were eager to turn his wit on himself in some show of surrender.

Over time the petulance, while never wholly gone from him, faded in frequency and intensity. Over time, as he told me his cautionary stories about life in the organization, he did so less with bitterness and more with self-deprecating humor. In these tales, never had he cast himself as the hero, but more and more I was invited to join my father in having a good laugh at the antihero's expense. Those stories, as I have said, had the effect of binding us closer with each telling. In the end, it proved no tragedy of alienation that my father would not, could not, share with me the details of his craft, the equations and accomplishments of his secret engineering at Lockheed. It was enough that he shared with me his regret about having chosen the work in the first place.

We are back on earth, my father and I. He has made a perfect landing at the Watsonville airstrip and we are sitting in the Mexican restaurant by the tarmac, tucking into two enormous burrito platters. As we do, my father tells a story, new to me, set in the last few years of his career.

He tells me that one day he received an order to take a polygraph test. This was routine for "people associated with certain kinds of projects," my father explains in his carefully vague way.

"You'd report to some outlying, nondescript building. The machine would be on a table with a chair beside it, in a little room with blank walls, and you'd be introduced to the operator. He'd strap you in, run a strap around your chest to measure breathing, attach fingertip pads to detect perspiration, put a sensor over your heart. Then he'd ask questions. Not about your lifestyle, but about the caretaking of classified materials. Have you ever taken classified documents home from work? Have you always stored your classified documents appropriately? Have you ever divulged classified information to a nonauthorized person?

"I think that last was the one I triggered the little squiggles on."

The polygraph test then became very much other than routine.

"I began to think of the conversations you and I had been having. I didn't really have this glittering recollection of how detailed things had gotten. Even acknowledging a black program existed was at one time not allowed. My mind wandered, and that's the wrong way to take a polygraph. With a wandering mind.

"There was also the matter of some notes I had been keeping which could be interpreted as classified information. Whenever you transport classified information, you're to double wrap it and carry it accompanied by another person cleared on your project. But on a few occasions I had carried those notes by myself, in a plain folder. I knew I'd been doing that. I acknowledged this. But this piqued the interest of the polygraph operator. 'If he's doing that, what else?' His questions kept getting more probing.

"I was called back for a second polygraph and I didn't do acceptably well on that one, apparently. Time goes by and I get a phone call from the local security office. 'You're to be in Washington Wednesday.' *Another* polygraph. So I got on an airplane and went out to Washington. In the morning of the interview day I was chatted up by a couple of men in the so-called Program Office. They said, "The issue is the transportation of classified information, so tell us about it.' I told them what I had done, that it was a matter of convenience with zero risk. They said, 'Everything's fine here. Have a good lunch. Relax. And be at a certain building at one o'clock.'

"But how can you relax? So I walked into this building, an ordinary office building with six people in the waiting room all sitting there, nobody speaking. I was ushered into this little room where I was introduced to the man who was going to be my host for the day. He was weird. He had this Bob Dole quality about him. He was trying to be friendly, but it just wasn't happening. You could tell that he felt he was in the presence of the enemy.

"He hooked me up and began to ask a lot of questions about the transportation of documents and the divulgence of classified information. It was then that I told of the chats you and I had had over the dining room table, that I *might* have said to you that I *might* have been part of a black program. This went on for a while. The needle would go scritch, scritch, scritch. He'd tear a sheet off the machine and say, 'I'll be right back.' There I'd be, strapped to this machine, staring at a blank wall. Except not all the walls were blank. One of them had a glass mirror and you knew someone was watching. It was like being in a monkey cage. I'd close my eyes, drum my fingers. Then he'd come back and ask another ten or fifteen questions.

"It began to get late. I had now spent three hours in this chair. I said, 'Hey, look, I've got an airplane I need to catch.' He said, 'You'd better change your reservation.' I went out to the receptionist, who was packing up for the day. I made a reservation for a late-night flight. The grilling went on. You begin to feel paranoid: 'They're going to find out things I didn't even *know* when I walked in here.' You'd spilled your guts to this guy, told him things you'd done, might have done, thought about doing, and he's still pursuing this. I began to think this was a career-ender. My security clearance would be revoked. I know of a man who blew his polygraph. He was frog marched out the door. I was thinking, 'My future could depend on what I've already got on this chart paper.'

"I don't remember whether or not he told me flat out that I did not look innocent. But finally the guy said, 'We're not getting anywhere here.' And then he asks me, 'How do you feel about all this?'

"I said, 'Just on the basis of what you've asked and how long it's gone on, I'm assuming the wiggles on the page don't exonerate me. But I can assure you, sir, that I'm a good soldier and the information has not fallen into the wrong hands.'

"He said, 'Ahhh, that's good stuff.' He shook my hand. 'Well, good luck,' he says, as if we were still good friends—just about as good as we were when I walked in.

"On my way home, driving to the airport, I resolved I was never going to take a polygraph again. It was so unpleasant and I was so poor at it, I knew that if I did nothing to violate security regulations, I'd still probably flunk the next one because I'd be so worried about flunking. I said to myself, 'I don't care if this is a condition of employment. I will refuse. I don't deserve it. I haven't earned it. They have no reason to suspect me.' So here was a case where the very mechanism used to identify the loyal employees was driving a dedicated and hidebound organization man like me right out of the organization. I was no longer a willing member of the group because of the way I was treated.

"What finally happened is a letter went into my security dossier, citing carelessness in handling classified material. It said, basically, 'We admonish you, and don't ever do it again.' I did not lose any access I already had.

"And why should I? I was too wimpy and scared to be a real threat. Look at this CIA guy who singlehandedly dismantled the U.S. spy network in the Soviet Union. He passed every polygraph because he had the skills. The people they catch are the honest ones. I'm the guy whose palms sweat!"

"Great numbers of professionals from many walks of life, trained to cooperate unfailingly, must be recruited. Such training will require years before each can fit his special ability into the pattern of the whole."

When Wernher von Braun projected this vision of how the new middle class would make its way in the world of work, when he spoke of this human system perfectly engineered so that his "flotilla" of spaceships would go to Mars while his ascendent tribe of technocrats cheered with pride on Earth, he did not explain how people could be made to cooperate unfailingly. He did not, for example, speak to the problem of the person who has special abilities to offer, yet because of gender or skin color or late middle age, is made to feel a poor fit. He did not say how such recruits might react to the news that they are no longer wanted within the pattern of the whole.

For much of the span of my father's career, there has been at Lockheed Missiles and Space Company a group of employees united in the accusation that they have been cheated of Wernher von Braun's promise. They have not found in Lockheed the meritocracy they sought, one ordered by objective measurements of skill and dedication, this data tallied reliably in each member's regular performance reviews. Instead, they see in their blue sky workplace a murk of prejudice and favoritism.

Members of the Lockheed Minority and Female Coalition, as they named themselves in 1970, are said to be "vocal" and "radical" in articles about them in the *San Jose Mercury News*. "I've covered a lot of labor disputes," a veteran reporter told me, "and this is the most militant employee group I have ever met."

The basis of their anger is to be found in the *Mercury*'s pages. In 1973 the coalition won a class action discrimination lawsuit, wringing from Lockheed a pledge to provide several thousand jobs to people other than white males. In 1985 a federal investigator's preliminary report, passed among employees, was "scathing," according to the *Mercury*, "full of examples of unchecked discriminatory practices and serious weaknesses in Lockheed's affirmative action efforts. A final report—after Lockheed had a chance to rebut allegations—was much more moderate but still critical." In 1991, Lockheed's own probe of one facility found it a bad place to be a female. "Any woman who is promoted or receives favorable personnel action is perceived to have had a romantic relationship with management," concluded the internal study. In 1992 an African-American named Norman Drake won nearly a million dollars in a lawsuit against Lockheed. He had been hired as a satellite engineer but demoted to security guard; along the way, he said, coworkers dumped garbage on his head, assaulted him, told racist jokes over the public address system, and hung up a picture of a black man in chains with "Norm, dumb nigger" scrawled on it. A white engineer named Johnny Atnip, claiming he stuck up for Drake and paid for it with the loss of his own job, settled his suit out of court. In 1993 a black satellite engineer named Charles Okoli was awarded $275,000 by a jury who did not agree with his claim that racism had thwarted his promotion at Lockheed. What the jury did find is that when Okoli cried racism, his superiors retaliated with "malice," branding him a "troublemaker" and inflicting other "emotional distress" upon him. In 1994, a new government assessment of the company's work culture found its way into the hands of employees, who turned it over to the *Mercury*. The front-page story began, "Lockheed Missiles & Space Co., which in recent years has faced a barrage of bias lawsuits, has discriminated against 109 minority job applicants and failed to meet a broad range of affirmative action obligations, according to a preliminary federal investigative report."

To all these episodes and to picketers charging racism and homophobia outside the company's gates, the response by Lockheed management has been consistent: The mechanism is sound but for an unavoidable glitch of human nature here and there, which is repaired as soon as the faultiness reveals itself. "There has been much more accountability put in the employment process, more training, more sensitivity at the upper levels, much more recruiting," Lockheed's lawyer told the *Mercury* in 1994. "There may be a fight or altercation [among employees]," company president John McMahon told the newspaper in 1993, but it was "dead wrong to say Lockheed is racist." Putting the blame on self-serving rhetoric by "a spattering of people who . . . certainly don't represent the community of employees at Lockheed," McMahon left the reader to take his implication. At a moment when juries were awarding huge sums in bias suits, a moment also when deep cuts were facing the community of employees of Lockheed, there was plenty of incentive to invent a victim's tale.

I heard the ring of plausibility, however, in the story offered me by John Farris, an engineer fifty-six years old with a neat mustache, wire-rimmed glasses, and black skin. The day we spoke, he had filed no lawsuit, had no air of militancy about him, yet he looked back on his Lockheed career as a promise broken all the same.

John Farris's voice kept the smokey burr of Oklahoma City, where his father had loaded cotton bales and butchered livestock. Reading a book on telemetry as a young man inspired him to get a degree in electronics and join Lockheed in 1965 as a missile electronics technician. Four years later he was promoted to publications engineer, writing research and documentation for missile tests. He enjoyed the work immensely for the personal initiative it demanded. "There was nothing holding me back." He moved his wife and three children into a ranch home in a south San Jose tract where all the streets were named for racehorses. Nights, he studied for a second engineering degree.

In the early 1970s a new supervisor's performance reviews rated him marginal for work John Farris considered well above average. That supervisor "almost dared me to complain," he said. "Like, 'What are you gonna do about it?' " Never use the word "racism" in such a situation, John Farris knew from the role-playing games in the management psychology class he was taking. To his supervisor, he quietly said, "You might have a preconceived idea about me because of what I look like." After that, the performance reviews and meetings about them worsened, until John Farris refused to sign off on one evaluation, saying, "If I had a worker like this, I'd fire him."

He was not fired, but was moved from one group to another and then others, his reputation preceding him each time. "I'm going to send you a problem," his new coworkers would be told, and so they would treat John Farris as trouble to be avoided. He was without a mentor in a place where mentorship had everything to do with who got the challenging work and its pay raises, said John Farris. Nor could he be a mentor, because when the boss was away, he never was placed in charge. His juniors kept receiving that honor. Then, when technical questions stumped them, they would say, "Go ask John." Slurs, the few times he heard them, tended to be posed as sly ribbing, like the time a colleague said to him, "If it hadn't been for Lincoln, maybe we wouldn't be having some of the problems we're having today."

After a while, John Farris sought escape from his "problem" reputation by working with subcontractors who knew merely that "I wore nice suits and drove a BMW." After a while, he stopped trying to tell his wife and children about the good engineering he did, assuming they would doubt him. "If you're not getting promoted, it doesn't look like you're telling the truth."

After a while, he began spending "too much" of his spare time on a fishing boat or communing with fellow Masons at the lodge, hours lost to his family while he was "trying to feel my manhood." His wife divorced him.

On an autumn day in 1994, he was told to see a manager who said, "Your number's up." John Farris could not tell me why he did not abandon Lockheed Missiles and Space Company before the company could abandon him. He wishes he had, because shortly after clearing out his desk he stepped on a plane to Ghana, finding there some interesting consulting opportunities for a black man with two engineering degrees. In that West African country, which he had always wanted to visit, John Farris found, too, a sense of "mutual respect" he had not known all those years in Lockheed's world of work.

For a brief while in the middle of the 1990s, nearly a quarter of a century after the first meeting of the Lockheed Female and Minority Coalition, new and different faces began showing up at gatherings of that "vocal" and "radical" group. They were white and male and fiftyish. They were victims, they claimed, of meritocracy subverted. They had spent years accumulating performance reviews and polygraph results until their files bulged with precise calibrations of their high worth to the organization. And yet that bulge in their files had become their enemy; Lockheed seemed to be firing costly senior workers in favor of youth, cheap and flexible. These rejected middle-aged white men were now in the process of gathering the data to prove themselves members of a definable, oppressed minority group. Lawyers had been hired, and when the data was in, they were going to file a class action suit for millions, just as other minority groups had their own class action suit in the works against Lockheed.

As it turned out, neither suit was filed. Lockheed management would offer this as proof that no basis ever existed; lawyers for the other side have told me that Lockheed sealed too many pertinent files in the name of national security, covering over injustice with the cloak of the black world. But when it came to proving age bias against white males, there were other problems, one of the lawyers explained to me. Many of his clients, some of the most angry ones, thought they could win the day simply by identifying the less worthy, those peers who still had their jobs just because they had played it more shrewdly, had made friends up the ladder. These clients were wrong in thinking that this, the way of all organizations, would move the court. Or, in the end, one of their own lawyers. "It's unpleasant to have to acknowledge that your own privilege as a white man helped you get as far as you did get," the lawyer mused to me. "And that is something I never heard from any of my clients."

He had never heard the stories of my father, of course, who freely acknowledges, "Many a time, when Lockheed was looking for people to toss out, I was rescued from the trash heap just because the right person liked me, a mentor stepped in and said, 'I'll take him in my program.' If I had been black or female or the wrong age, that wouldn't have happened." When my father retired, the people of his program held a lunch hour party with a cake and coffee and speeches. He was by then high enough in the hierarchy to have negotiated and guided projects worth $100 million a year. He was valued, clearly, by the forty well-wishers at his going-away party, who gave him a statuette of a winged dragon grasping a jewel in its talons, with the inscription, "To HAL BEERS In Recognition of 32 Years at Lockheed Providing Leadership and Dedication To Programs of National Importance." The dragon with jewel was symbolic, apparently, of whatever mysterious system my father had helped engineer in his last years at Lockheed.

My father brought home the statuette, placing it on the desk where he pays bills and changes the diapers of his grandchildren when they visit. On and around the desk, he arranged other items. A plaque: "EAGLE AWARD Presented to Hal Beers on this 24th day of September, 1992. For Vigilant Guidance Provided to the Subcontractor from the Matrix Subcontractor

Team.'' A pen and pencil set: ''In recognition of the significant contribution to a national program essential for the security of the United States of America.'' A plaque: ''Hal Beers'' on a brass plate below a picture of a ghost emerging from an opened safe.

These mementos, I thought when I saw them, were hints of those stories my father never could tell me, stories that might explain why he does declare himself, today, to have lived the life of a lucky man. Forever bound by his oath of secrecy, my father would offer only this explanation: ''I've been fortunate enough that on more than one occasion, our accomplishments have been very gratifying and technically extraordinary.'' Whatever miseries the organization had inflicted, whatever sacrifice the blue sky tribe had exacted, he was lucky to have what the unlucky ones did not. At the end he was given cake and speeches, a dragon statuette and modest affluence for his remaining days, all of it proof of receipt of one life tendered. He was given reason to know he had fit well within the pattern of the whole.

I phoned Gary Kolegraff one August evening, six months after our last conversation, to see how he was faring under the rules of the new economy. Quite well, he assured me, the chipper voice turned exultant. He'd been hired by a small, aggressive company that used satellites to map routes for police and other drivers in a hurry. He was paid better than at Lockheed, and this where the average employee was twenty-five and the chief executive officer wore Bermuda shorts. He had landed this accounting job not through the Career Transition Center of Lockheed, nor through the intercessions of Queen of Apostles, but by walking through the front door of Navigation Technologies and asking for an opportunity to prove himself. And then, having avoided the dreaded move home to Mother and Father, he had boarded an ocean liner to the Virgin Islands. These days, he was back in the market for a girlfriend, perhaps marriage and a family of his own.

''I've got this lady I met on the cruise coming out to visit.

She's from Greenwood, Indiana. I think she's a secretary; I'm not positive. We'll do the Monterey thing, drive around. If that works out, we'll have to see how it goes. I don't want to spoil her too bad!"

What remained of his Lockheed severance package he had invested in the stock market, buying into Microsoft and betting on a firm that made ozone-safe refrigerants. He was developing a taste for risk, finding it invigorating.

"I probably have more shares than I should. But I love it. I like taking a chance. My roommate, the one who still works for Lockheed, he can tell by my mood that I'm happy. I would never, ever go back there! Deep down he knows he should leave. But he's in that phase I was in. I feel like shaking him and telling him, 'Hey! There's a whole world out there!' "

Risk, very small but real, is what compelled my father to stop flying those many years ago. Should he die in one of the small plane crashes we occasionally would read about in the newspaper, we would be left without a father and his income. It would not be fair to expose us to such risk, he believed, nor the risk of his trying to reinvent himself merely because he was bored and dissatisfied with his work at Lockheed. He would live with his choices of work and not risk the loss of security, the loss of stability, that changing career in midstream entailed. Plan the flight and fly the plan, the Navy had taught my father, and that is how he would proceed with his life.

My father is explaining this to me as he maneuvers the Cessna 172 out of a slow banking turn, the silver blue of the Monterey Bay slipping behind the mountains at our backs, Silicon Valley's labyrinth of lanes and cul-de-sacs presenting itself before us. My father tells me that when he was just out of the Navy, he had expectations of joining the exciting and select society of engineering test pilots. But certain events overtook him: marriage, me, the job offer from Lockheed, California, three more children. There was one instance, he confides, when he did leave his desk at Lockheed and make inquiries about how a man like him might change his career and become a working aviator again. But neither the world of test flying nor airline piloting had any use for my father because somehow, without his marking the time closely enough, he had become thirty-seven years old.

This story of a foiled, last-ditch escape attempt is new and surprising to me and it prompts me later, as we drive along the World's Most Beautiful Freeway in my father's brand-new, silver and streamlined Camry, to ask him for one more revelation. I have never, in the twenty years after my father beat my face black and blue on a summer evening, summoned that memory from him. I have long since forgiven him for it, have long since returned the crush of his hugs with love to match, but I am wanting to know the answer to this remaining mystery in my father. So cataclysmic an episode must hold a key, certainly, to the puzzle my father was then. The car is quiet and warm, and my father and I have had a good day in the sky, and so I ask him, "Do you remember it?"

"Yes," he says, "I hit you, didn't I?" He wears the expression of one asked to recall the plot of an obscure movie. By the look on his face, I presume he would prefer to forget. I feel bad for bringing it up and I do not expect any more revelations today. Then my father says, "I have been abusive to every one of you in the family. There are *so many* bad times I've lived to regret.

"Back then, I found myself in a very disturbed state of mind for days on end. And then I'd snap out of it. I was in my late thirties and my life was crystallizing. I could clearly see that things weren't going to turn out the way I had hoped, the way I had expected when I felt I had plenty of options. I began to feel hemmed in, penned in.

"None of the benefits, none of the very powerful pluses that were coming my way on a daily basis for having married Terry and for having you and your brother and sisters in the house, none of that seemed to make a bit of difference. When that was all, really, that I should have ever cared about. I was pissed off all the time and life looked to me a very dreary landscape. And what that leads to is, if you're disappointed in yourself you find yourself doing bizarre things that hurt the people you love. You pick on the people you love and the reason you do it is you know they'll forgive you. And that's what I would do. Even today I fly into little rages with the people I love. I hope they are fewer and less severe and not as long lasting. It's just a phenomenon I learned about myself, and I see it for what it is."

My father tells me then about how he came to see this phenomenon for what it is. He would rage at his wife or child and the next morning he would arrive at Lockheed sick with remorse that ached like an alcoholic's hangover. He would sit, then, at his desk writing what coworkers assumed was documentation related to a secret project of technology. In truth, my father spent hour upon hour writing long essays to himself, recording feelings observed within, working his way painstakingly toward some diagnosis of the irrational at his core. Whatever came to mind he wrote down, blame heaped not only on himself for becoming a "paper pusher," but also on his wife and children for blocking his way out. What he wrote was often raw and ugly and he never meant it for other eyes, my father tells me. But he kept every page in a folder by his desk, a folder that grew thick over the course of fifteen years, and at times when the hangover of contrition had lifted, he would pull old essays from his file to see where he had been and whether he might now know enough to troubleshoot the problem.

What my father concluded is this. "Puzzling over why I was such a bizarre personality, I resolved it was the difference between myself as reality displayed itself to me, and the inflated self-image I carried around." My father's task, the task that Lockheed Missiles and Space Company had in a sense assigned him, was to sit at his desk and write memos to himself until the expectations of a hot young fighter pilot had been reduced to fit the space that had in fact been reserved for him. When my father wrote his last letter to himself, sometime in the early 1980s, he passed the entire file through the shredding machine Lockheed provides for classified documents of no further use.

I love my father all the more for telling me this story, for adding to his familiar theme of regret the honest admission that we, his wife and children, had not only given him much, but had exacted a cost as well. I thank him for the doubt and pain he has revealed to me, layer by layer, not just through his stories but even when I knew it as inchoate anger. I am grateful because all of it showed me why the culture of the corporate bureaucracy was a way of work not only to be avoided, but unlikely to thrive forever. Those organization men and women who shared his misgivings but repressed and ignored them, choosing instead to force a happy face at the dinner table every night, did their children a disservice. If my father had not exposed for me the flaws in its foundation, would I have managed to be so far clear of the blue sky monolith when the toppling began?

On the freeway crowded with commuters finishing their day of work, my father finds a slot in the formation between a rusted Dodge pickup carrying Mexicans and gardening equipment, and a Mercedes sedan with a license plate boasting ABUV PAR. I turn the conversation away from disappointments, toward the sky. I ask my father why the former jet pilot had learned to fly powerless sailplanes years ago, just before he quit flying altogether.

"Soaring is a battle of wits," he says. "You are an intellect dealing with the airmass. It's invisible but you are trying to exploit its properties in a way that favors you. And if you don't do it well, the negative aspects of that airmass are going to force you to land sooner than you wanted to.

"So there's a case where, sure, you're encased in a machine, but really what's going on is intellectual."

Gary Kolegraff cheerfully tells me that his Navigation Technologies job did not last past three months, that he has landed and lost many other jobs after that, and that he is, for the moment, without employment.

"These companies are generations ahead of Lockheed and I find it a really hard fit. They actually make you work! Believe it or not, I may go back to Lockheed for a temp job."

But not if his latest dream were to come true. He has taken the savings in his Lockheed 401K and rolled it into high-risk, high-growth investments, some of which, like a software maker for the Net, are yielding extremely well. His goal is to clear a $100,000 gain in one day and, if all goes well, to be a millionaire in a few years. Then he will never need a job at all.

"The other thing I've been doing is hanging out in coffee bars. That's what I enjoy most. Just having a cup of coffee and taking a hike all day."

The cruise ship girlfriend from Greenwood, Indiana, had not been his type, which is fine because he has decided he lacks interest in marriage, anyway. He is living, still, in the rented tract home with the same two single men. Near as I can see, Gary Kolegraff has reinvented himself into a component very attractive to the new economy: the white male unburdened by family, accustomed to the notion of no job (much less a lifetime career), the coffee shop dweller ready to bet his retirement fund on the rises and falls of virtual corporations making their runs at the global market.

"I got my father interested in investing, too. He's making a fortune in tobacco."

I hang up the phone and move toward the sound of my baby daughter crying. We are alone in the condominium, Nora and I, for this is one of those days her mother spends at the university where she is a professor of education. I meet Nora's reaching arms with my own, lifting her from her crib and pulling her tightly to me as we go to the diaper-changing table. At the advanced age of thirty-eight, I am enclosing a child of my own, finally, in a father's hug.

My wife and I are every bit the cliché of our time as my parents are of theirs. Like any truly modern couple, we do not cede separate realms to each other as did my engineer father and mystic mother. We have been soulmates since our meeting in college, Deirdre and I, drawn together by all we have in common. We are Irish Catholic children of sunny suburbs who share ironic agnosticism, leftward politics, a taste for impolitic humor. We make our livings with words on computer screens and are always interested to read whatever latest product the other has manufactured from those words.

The conceit of the son of the suburbs who moves to the city is that his choice is bolder than the example of the parents. But I don't think that about myself anymore. Mine has been the more timidly conservative life, in many ways. I did not gamble my happiness on four children before I was thirty-five, nor was I audacious enough to believe I could invent meaning in the vacuum of a freshly sterile subdivision. Deirdre and I have chosen to live in a city of beauty and every amenity, a very easy place to be with no child or even with one. So flush with Asian investment is Vancouver, so perfectly manicured are its parkways and neighborhoods and tourist-friendly attractions, that this city has become the highly desirable suburb to the Pacific Rim. When Diefenback Elkins, a U.S. design firm, advised Air Canada to market all of Canada as "a kind of innocent America, as yet untainted by ethnic tensions and urban blight," the target customer was me. A blue sky child is raised to recognize the fragrance of optimism in the air, and, for now, Vancouver is heady with it.

From Vancouver we ship to Nora's grandparents packages stuffed with pictures and videotapes of baby firsts. In return we receive clothes, toys, safety-approved car seats, and Jolly Jumpers, all the equipment necessary for raising a modern child. Clearly my mother and father, like Deirdre's parents, can hardly contain their relief that Nora exists, that we have made this demonstration of faith in the future. It does not seem to matter to them that Nora will not have what they gave me. She will be raised in the vertical, lawnless downtown, without a cul-de-sac full of playmates or a walnut tree to climb. She will not have a restless, fix-anything force of nature for a father, nor will she be shown the workings of my mother's heaven. Her father will not be one who planned his flight and now flies his plan, for I have no plan. I have, merely, a nervous sense of the shifting movements of air that we, and now Nora, too, float upon.

EPILOGUE

VIRTUAL BLUE SKY

Once you're cleared for take-off, we'll initiate your catapult down the launch tube. Now remember, you'll be at full throttle by now. Once you're movin', give the stick a little backward pull and get your nose up. And you'll get your lift from the end of the deck. . . .

The man telling me this, telling me how I will soon fly a jet off the deck of an aircraft carrier, is jut jawed and tan faced, wearing the familiar brown cotton and silver bars of a Naval flight commanding officer. He is explaining to me the procedures for operating the X-21 Hornet, "a top secret, high performance strike fighter that just rolled out of Strategic Command R&D labs." The genius of this machine, the evolved result of a half century of American aerospace, is that a person like me, a jet pilot's son who never learned to fly, can pilot the X-21 Hornet with less than an hour of training. That is why I am here in this windowless briefing room on Spacepark Way, preparing myself to take the controls.

. . . *You're gonna learn real fast why they call it a joystick. It's a state-of-the-art pressure sensitive mechanism that requires the slightest force in any direction to get a big, and I mean big response. . . . The Hornet is the Defense Department's answer to the security of our nation in the 21st century, with the most sophisticated aerodynamics, the most advanced avionics, and the most accurate weapons in the world . . .*

In this script (and it is a script, for my "Commanding Officer" is "Max Power," an actor speaking to me through a television screen), the precise reserve of my father's aviator language devolves into corny macho-isms I recognize from elsewhere, Clint Eastwood movies and Tom Clancy novels and Desert Storm headlines in *Newsweek*. These are glitches that break the spell, collapse the whole effect, but still, I tell myself, this is experimental stuff and I am not interested in waiting until the product has been perfectly engineered.

I have paid my twelve dollars and seventy-five cents to Magic Edge, Inc., was issued a flight suit at the cash register, and had my new code name, Astro, recorded in the computer. I have been introduced to my squadron leader, a live person, clean cut and all business as squadron leaders must be. He has collected me and three other customers, men who have strolled over from one of the tilt-up office buildings across the street in search of an airborne coffee break. We instant wing mates have been pounded with the tape-recorded thunder of fighter jets passing overhead as we have followed our squadron leader down diamond-plate steel stairs, beneath gray-girdered bulkheads, into the briefing room where Max Power lives on laser disk.

. . . *All right, people, I've heard you're a bunch of hot shots, and I'm anxious to see if the rumors are true. I'll be keepin' an eye on you and I hope you like this dream machine as much as I do. Good luck and good flying.* . . .

My wing mates and I are strapped into the cockpits of our respective X-21 Hornets, the black Plexiglas shells closing over us, the control panels glowing before us, the joysticks in our hands. I hear the voice of my squadron leader in my ear, telling me to throttle up. I feel the jolt of the catapult launch, the surge of acceleration into the simulated blue sky ahead.

Now I am high above the California of Don Morris, the baby boomer who dreamed up the X-21. His is the California where, in the spring of 1995, the entertainment industry eclipsed aerospace as the state's largest employer. His California is Disneyland, an invented place that, Morris has told me, he yearned to experience while growing up in the Midwest. When he reached Northern California as a young engineer at Hewlett-Packard, he loved to jet south for a day and ride Disneyland's Star Tours, "my first experience inside a real simulator." As a young father, he fashioned a two-seat prototype of the X-21 in his suburban garage, blocking the gap under the door to the kitchen with towels to keep the Fiberglas fumes from seeping in. When the home project was done, he and his son would sit side by side inside their plastic capsule, immersed in a *Star Wars* movie playing on the windscreen before them. What remained was to make the prototype flyable, the windscreen interactive. This was achieved by adding hydraulic rockers to buck and roll the pod, and by flashing before the pilot software imagery by Paradigm Simulation Inc., a maker of flight simulators for Boeing, Lockheed, and other military contractors.

At this altitude above virtual California, the effects of the disappearance of more than half a million aerospace-related jobs are invisible to the eye. Swooping lower offers no better understanding. In my neighborhood the saplings lashed to stakes have grown into trees, there are far fewer children in the street, and Shopwell is now Yoahan, an all-Japanese store selling rice, seaweed, and pickled oddities in bulk. But there are no husks of factories or rows of boarded-up houses to inspire folksongs about hard luck aeronautical engineers. Where blue sky culture fades, the suburban veneer remains implacably pastel.

A hard pull back on the joystick aims my X-21 higher and higher, until the virtual blue tints with the ink of space, a blackness that is, these days, returning to what it represented before *Sputnik.* During that strange decade or so after *Sputnik,* astronauts were seen to be blazing a trail so that any citizen might one day make a personal visit to the place of space. No one really believes that anymore. Better to see in the blackness of space a vast projection screen, the use film director Fritz Lang made of space early in this century with his *Girl in the Moon.* Better to inhabit the space within our video monitors, where personal fantasies of individual freedom and morphing identity can be lived. Hostile outer space, an airless freeze where survival demands total obedience to "procedure," could never offer that. What, then, has a national space program to offer the imagination of me today, as I pilot my X-21 dream machine? Amid the hundred thousand pieces of space junk that swirl around Earth, the authentic astronauts try to catch a satellite on a pole or reel one out on a long tether. As Don Morris and his son so well understood, even an old George Lucas movie is far more entertaining.

Or better still, one of Captain Kirk's alien romances. My father's industry has gone the way of *Star Trek:* Originally watched to see how the Manifest Destiny of the American way would play out, it is now scavenged for kitsch. The famous footage of Neil Armstrong hoisting the American flag on the surface of the moon is replayed on MTV, albeit with an MTV flag. The hull of a rocket used to launch a satellite advertises Schwarzenegger's *Last Action Hero.*

Small wonder that the last president's call for a $400 billion mission to Mars met quiet coughs, that the space station keeps shrinking, that due to lack of interest we will not land a probe on a comet after all. No, the only thing a national space program has to offer me, pilot of the X-21, is what the Hubble telescope so efficiently mines from the galaxies: images. Once they have been properly computer enhanced and digitized, I will be able to fly through those sunset thunderheads Hubble has shown us, those pillars of star gas seven thousand light-years from Earth and trillions of miles high, as beautiful in their chiaroscuro rendering as anything that Chesley Bonestell ever painted for Wernher von Braun.

I nose earthward, locking in bearings for the plains of Kazakhstan, leveling out for a low flyover once I have in my sights the Baikonur Space Center of the former Soviet Union. There the most powerful rocket engines ever built, two twenty-story Energiyas, lie ten years dormant in a hangar, feathers and bird dung coating the platforms around them. "If you could just find some rich guys to pay for them, I know we could send them up." That is what the chief engineer of the space center told a reporter for the *New York Times*.

Space City, the housing development built for Baikonur workers, is a place of breadlines nowadays, a society without enough milk for its children. Of what work remains, some is due to the fact that my father's employer lost a contract, in a sense, to the Russians. The design of the space station's escape vehicle, once a Lockheed program, is now in the hands of Russian rocket scientists who are paid four dollars an hour. I know this because I spoke with a Lockheed engineer just back from a consultation with his former adversaries. He was an old hand who considered this his last project. What he remembered most was the feel under his feet of the stairs of the Russians' design complex, the saddle of wear in concrete left by aerospace engineers much like him (except for being the enemy) as they went about their millions upon millions of work days.

The trigger on the Hornet stick will fire off a motherlode of munitions. . . . I gotta hand it to those R and D guys because the weapons onboard are sweet. . . . *My personal favorite: your high velocity projectiles, referred to as HPVs. Now, these babies are kinda like skippin' rocks at mach four. . . .*

With memories of Max Power's briefing singing me along, I veer southwest, skim the Caspian Sea, Tehran, Baghdad. Desert

Storm, that booster of consumer interest in the product I am flying today, was a Cold War's worth of aerospace systems engineering for all to see: the satellites and drones and AWAC planes mapping the attacks; the Tomahawk cruise missiles and A6E Intruders and F-18s and F111s and F-4G Wild Weasels and A-10 Tankbusters and F-15E Eagles and B-52s and Tornado GR-1s and Apache helicopters and F-117A Stealth strikers wreaking their various specialties of destruction; the KC-135 and KC-10 air tankers circling above for midair refueling; a mere eight allied planes crashing to Earth in the first two thousand sorties. After the nasty, illicitly thrilling shape of the Stealth fighter, what icon of power is left for aerospace to put in the sky? After the traffic jam of retreating Iraqi soldiers was turned to a smear of immolated bodies on the sixty-mile stretch of road out of Jahra, Kuwait, what gesture of technological omnipotence is needed?

In a blink of the imagination, the East Coast of the United States rolls into view, and then the factory towns and farmhouses, the industrial parks and trailer parks, the America that lacks focus when subjected to the scrutiny of focus group facilitators. I am flying over the fifteen American communities where, at the end of 1995, focus group discussions were conducted for the Pew Center for Civic Journalism. The man who summarized the findings, Richard Harwood, found Americans believing their nation was no longer a fair place to work, believing their economy was "unraveling before their eyes," believing that no institution—government, corporations, the media—reflected their concerns. "People are deeply ambivalent about what we should do. They believe this nation is entering a new era. They're not looking to return to the '50s. They know that's not desirable. But they don't believe that we have the capacity to work through these major shifts."

Finally a familiar circuitry of Silicon Valley freeways appears before me, and then all the cement-walled tilt-ups surrounding the cement-walled tilt-up that houses the Magic Edge, Inc. entertainment center. It would be incorrect to consider the complex nothing more than one more elaborate video game parlor. Those are intended as places for teenagers to spend their quarters consuming electronic "thumb candy." But Don Morris and his cofounders of Magic Edge, Inc. saw a very different market for their product. They set out to create a virtual *culture* consumable by adults in twelve dollar and seventy-five cent increments. Don Morris has told me his product is geared for people with good jobs, the kinds of people who work in the high-tech business campuses surrounding his complex. Should they want to slip into a flight suit uniform and hear Max Power extol their worth to the organization, Magic Edge, Inc. will sell them what they desire. The public relations person at Magic Edge, Inc. assures me the center is drawing thousands of customers a month. Especially popular are package deals offered to corporations who see in the X-21 a tool for employee "team building."

My allotted eight minutes in the X-21 Hornet are coming to a close. I realize I have done little other than settle in at high altitude and let my mind do the aerobatics. The simulation software shows me an image of the Golden Gate Bridge, and I think to myself that it would be exciting to fly under the span, a trick my father could only have dreamed of pulling off in his Grumman F9F-8 Cougar way back in the 1950s. I ease the joystick forward, apply the air brakes some, begin to line up my approach.

But the X-21 is suddenly in free fall, the hydraulic rockers beneath pitching me forward, the pixilated azure of the Pacific Ocean filling more and more of my windscreen.

This is what comes from not paying attention during the briefing. I had not recorded the fact that this day's other fliers, my squadron mates, were not only allowed but invited to accumulate extra points by shooting down one of their own. "You are hit," the voice of our squadron leader informs me in my headset. The controls are useless in my hands, and virtual blue is all I see.